# GRAIN DRYING, HANDLING, AND STORAGE HANDBOOK

MWPS-13 THIRD EDITION

*Dirk Maier*
*Samuel McNeill*
*Kenneth Hellevang*
*Kingsly Ambrose*
*Klein Ileleji*
*Carol Jones*
*Mark Purschwitz*

Iowa State University / Ames, Iowa

MWPS-13



Book design and production by Kathy J. Walker

Third edition, 2017

Library of Congress Cataloging-in-Publication Data
Names: Maier, Dirk (Dirk E.), author. | Midwest Plan Service, issuing body.
Title: Grain drying, handling, and storage handbook / Dirk Maier, Samuel McNeill, Kenneth Hellevang, Kingsly Ambrose, Klein Ileleji, Carol Jones, Mark Purschwitz.
Description: Third edition. | Ames, Iowa : MidWest Plan Service, Iowa State University, 2017. | "MWPS-13." | Includes bibliographical references and index.
Identifiers: LCCN 2017029322 | ISBN 0893731102 (pbk. : alk. paper)
Subjects: LCSH: Grain--Handling--Handbooks, manuals, etc. | Grain--Drying--Handbooks, manuals, etc. | Grain--Storage--Handbooks, manuals, etc.
Classification: LCC SB189.7 .M25 2017 | DDC 633.1/0468--dc23 LC record available at https://lccn.loc.gov/2017029322



MWPS would like to thank the following reviewers for providing support to this project:

CHAPTER REVIEWERS
Tom Brumm, Iowa State University, Chapter 5
Carlos Campabadal, Kansas State University, Chapter 3
Michael Goetz, Gateway Building Systems, Chapter 2
John Lawrence, Kansas State University, Chapter 7
Ron Noyes, Grain Storage Engineering, LLC, Chapters 3, 4, and 5
Randy Poock, Sukup Manufacturing Co., Chapter 4
Charles Schwab, Iowa State University, Chapter 6
Steve Waechter, Sukup Manufacturing Co., Chapter 4
Gary Woodruff, The GSI Group, Chapters 1 and 5
Darren Zink, Brock Grain Systems, Chapter 2

BOOK REVIEWERS
Dirk Maier, Iowa State University
Kenneth Hellevang, North Dakota State University
Carl Bern, Iowa State University

FUNDING
This edition was in part funded by a grant from The Andersons Inc. (Research Grant Program) administered by The Ohio State University through the NC-213 Multi-state collaborative research project.

MWPS would like to thank the following people for providing support to this project:
John Moore, Editor, Biddeford, ME
Letitia Wetterauer, Illustrator, Alpine, TX

DISCLAIMER
Mention of a trademark, propriety product, or commercial firm in text, tables, or figures does not constitute an endorsement by Iowa State University or the publisher and does not imply approval to the exclusion of other suitable products or firms.

# CONTENTS

# FOREWORD

This third edition of MWPS-13: *Grain Drying, Handling, and Storage Handbook* is a substantially updated version of the previous edition. This new edition was prepared by a team of agricultural engineers who have extensive expertise in engineering post-harvest systems along with real-world experience in planning, constructing, and operating systems for handling, drying, and storing grain. Members of this team also have broad experience training university students, farmers, and grain elevator managers in operating these systems to successfully preserve grain quantity and quality.

The authors readily acknowledge the authors of the previous two editions—Leslie Backer, North Dakota State University; Roger Brook, Michigan State University; Harold Cloud, University of Minnesota; Kenneth Hellevang, North Dakota State University; Bruce McKenzie, Purdue University; William Peterson, University of Illinois; Rich Pierce, University of Nebraska; Larry Van Fossen, Iowa State University; and Gerald Riskowski, MidWest Plan Service Engineer. Several of the current authors have had the privilege to be part of the second edition and to have known authors of both prior editions. They were truly pioneers in developing the original designs and plans for on-farm grain drying, handling, and storage systems at a time when mechanization rapidly increased the volume of harvested crops. Off-farm grain facilities were not able to keep up with harvest capacity and with the on-farm demand for grain to feed livestock and poultry. The authors of the previous editions of this handbook worked to solve these problems, and their efforts laid the foundation for many of the ideas more fully developed in this third edition.

For a number of years now, farm operations have been growing and the demand for larger drying, handling, and storage capacities has been increasing. Thus, this third edition is a timely contribution in support of this trend. The handbook's content has been substantially updated and expanded to reflect dealing with the larger grain volumes that accumulate at on- and off-farm facilities more rapidly during harvest than ever before. The need and importance of safety and automation that comes with faster and larger facilities is reflected in the addition of two new chapters; Chapter 6, Safety and Health in Grain Drying, Handling, and Storage, and Chapter 7, Automation in Grain Drying, Handling and Storage.

We thank our coauthors for their willingness to contribute their expertise, knowledge, and time, without which this handbook would not have been possible. We also thank the peer reviewers of these handbook chapters. Their substantive and timely reviews of chapter drafts improved the content and relevance of the entire handbook.

Thanks also to the staff of the MWPS office at Iowa State University—Dr. Jay Harmon, Professor-in-charge, and Ms. Kathy Walker, Program Assistant, for working patiently with us on this project for longer than originally planned. Ms. Walker's ability to keep this project moving forward, along with her due diligence in pulling together and formatting all chapters including figures and tables, has resulted in the high quality of this publication. Without her, this project would not have seen the light of day.

Last but not least, we express our sincere appreciation for the financial support of the Andersons Research Grant Program, which is administered by The Ohio State University through the NC-213 Multi-State collaborative research project. Their funding provided much needed support for preparing, editing, and printing this new edition.

Most importantly, this handbook is for you—farmers, elevator managers, equipment dealers, students, and others interested in developing grain systems. The editors trust it will meet your needs and exceed your expectations. They welcome your feedback and suggestions for future improvements. Also available are on-line resources, such as spreadsheets, that complement the content of this handbook. They are available at www-mwps.sws.iastate.edu/spreadsheets.

CHAPTER 1

# **PLANNING** GUIDELINES

*Dirk E. Maier, Ph.D., P.E., Iowa State University*

This handbook can help you make good decisions while planning on- and off-farm grain handling, drying, and storage systems. Planning involves **selecting** specific pieces of equipment, **sizing** their capacities, **organizing** them into a functional system, and then properly **locating** the system.

The importance of planning cannot be overemphasized. It requires considerable time, effort, and thought. Good planning results in a system that works well now and can expand for future needs. A well-planned system usually costs no more than a poorly planned one. This handbook is for farmers, grain elevator managers, equipment dealers, and others interested in developing grain systems. It applies to all grains, oilseeds, and pulses which we refer to as "grain". Recommendations are based on research, when available. There is more information on corn than on other grains, because more research has been done on corn. Where research is lacking, guidelines are from the experience and judgment of agricultural engineers representing the leading Midwestern Land Grant Universities as well as industry. Also, the perspective distinctly represents the U.S. Midwestern corn and soybean region, where on-farm storage capacity has traditionally been about twice as large as commercial grain storage. However, most, if not all, of these planning guidelines apply to any region in the world where on-farm and smaller-to-medium size off-farm grain facilities are being considered.

Read the entire handbook to understand the whys and hows of developing a total system for grain drying, handling, and storage. Then re-read, study, and apply the sections that interest you most and that may help you improve and expand the layout and operation of your facility. You may also want to consult experienced engineers, knowledgeable equipment suppliers, and grain producers and elevator managers with systems similar to the one you are planning.

*Although they are important concepts, this handbook does not provide detailed discussion of the following subjects:*

- **Operation and management.** For operating and management techniques not discussed, consult owner's manuals, local extension educators, state extension agricultural engineers, and on-line resources such as those found at https://www.ag.ndsu.edu/graindrying and http://www.extension.iastate.edu/grain/.
- **Capacity of harvesting equipment.** The capacity of harvesting equipment is discussed only for its effect on sizing an operation's capacity to receive grain.
- **Transportation.** Transporting grain from field to storage and from storage to market or feedlot is mentioned only for locating the system or sizing equipment.
- **Feed processing.** Although you need to consider how grinding, weighing, and mixing fit into the grain system, this handbook does not discuss feed processing.
- **Handling high-moisture grain.** High-moisture grain and wet distillers' grains with solubles (WDGS) are excellent for livestock feed but have limited storage life. Consider equipment needed to include handling and storage of these materials in a dry-grain system.
- **Economics.** Before you decide to proceed with a project, and well before construction begins,

evaluate the economics of the plan. Your goal is to achieve the most functional system that is also economically feasible for you.

## PLANNING FOR SAFETY

An important goal of the planning process is to develop safe facilities. Chapter 6 of this handbook discusses fully the safety and health concerns involved with drying, handling, and storing grain. However, good planning can help establish a culture of safety and provide a comprehensive strategy for dealing with safety issues. Unsafe facilities are unacceptable no matter how functional or economical they may appear. Develop your grain system with the goal of protecting everyone—owner, operator, employees, visitors, family members, and especially children. Obtain safety awareness and educational materials from your insurance provider, local county extension office, or state safety extension specialist, or on-line such as at http://grainsafety.org. Use these materials to talk with your family, coworkers, employees, friends, and service providers about safe practices for handling grain. During planning, emphasize awareness and knowledge of hazards and risky operating procedures. Plan to minimize the dangers associated with common hazards such as PTO-driven shafts, pulleys, chains, conveyors, falls, entrapments, large vehicles, dust, and noise and electric shock.

Recognize that automatic equipment that can start unexpectedly is especially hazardous. Plan your layout accordingly, and post signs on automatic equipment warning that it may start at any time. If you are remodeling an existing facility, plan to install protective shields on operating equipment, and replace missing ones with new ones. Install lock-out and tag-out devices on electrical service to equipment for unloading storage structures(s). Also, plan for the capability to lock-out and tag-out operating switches and structure entrances whenever equipment is being serviced.

## LONG-RANGE PLANNING

Farm and grain elevator structures have a long life—at least 20 years. The productive life of a farm or elevator manager is perhaps 40 years. So you may have only two chances (every 20 to 25 years) to correct a planning mistake. Take seriously the process of long-range planning. Planning principles for large and small, on- and off-farm systems are similar. The main differences are grain flowrates, drying capacity, storage capacity, and automation level. With a young family, a group of young employees, and limited resources, it makes sense to keep an operation simple and rely more on manual work. But, when you are 10 years older and 20 years more tired, you may want additional mechanization and automation. Ask yourself whether installing more mechanization and automation is a wise choice now, and determine whether such features can easily be added later.

Grain systems include equipment for handling, holding, drying, cooling, and storing grain along with facilities that must operate as a coordinated and functional unit. The goal is to receive, dry, and transfer grain to storage or loadout, while maintaining quality and quantity without impeding grain harvesting or shipment. You may need to change your grain system to include increased storage, a higher-capacity bucket elevator, dust control, pneumatic conveying, wet grain receiving, a new drying method, a switch to livestock or cash grain, or feed processing.

Incorporate present and future needs into a long-range plan with provisions for growth and improvement. First, evaluate present needs; then consider possible growth and change for 3 to 5 years into the future. Then **double the volume of grain handled** and **re-plan** for this expanded need. The idea is not ridiculous; not everyone will double their production, but doubling the plan forces you to evaluate what might change within the facility's probable life.

Most well-designed grain and livestock facilities can be doubled and still perform well. If you plan for 3 to 5 years into the future, a doubled unit may serve 6 to 10 years. If the doubled plan is good, it likely can be doubled again without serious compromise and be useful up to 20 years, which is a reasonable planning base for most building facilities.

Collect company literature and equipment specifications. The equipment must handle present needs and fit future plans. While deciding on the type of system you need, visit farms and grain elevators with different systems and equipment. Look at alternatives, advantages, and disadvantages. After selecting a system type, visit systems of comparable size with similar equipment. Ask the operator what works best, what causes problems, what changes would improve operations, and what equipment needs the most repairs. In addition to the operator, talk to others who may be candid: a partner, spouse, son, daughter, employee, or neighbor.

Consider the effect of changes in grain or livestock volume on timeliness, capacity to get the job done, grain quality, and dependability of services (for example, custom harvesting and hired help). Plan for expected volumes of each grain produced and to be handled. Analyze the system for changes in grain production, feeding versus marketing, and other operational changes.

Though larger storage bins generally cost less per bushel, smaller bins add flexibility for different grain types or grain qualities. Cropping patterns may change; the identity of grains may need to be preserved, or some grain may be moldy at harvest, so consider at least two bins for each type of grain to add flexibility.

Select the general types of equipment and sketch the system. Sketches are easy to change, but a concrete foundation is hard to move. See Chapter 2 of this handbook for more information about site selection, arrangement, sketching, and system evaluation.

## PLANNING FOR GRAIN QUALITY

Your plan should consider facilities that will allow you to maintain high capacities and grain quality while minimizing damage caused by drying, handling, and storing. Grain quality at harvest is influenced by grain type, weather, and harvester adjustment and operation. Minimizing grain damage, especially to corn, to maintain grain quality requires good drying, cooling, and handling equipment, and conscientious stored grain best management practices.

*High quality grain*

- Has few fines and foreign material.
- Has little physical damage and few stress cracks.
- Has little mold and insect damage.

Clean grain dries and stores better. Plan to install a grain cleaner to remove fines and foreign material or at least plan space for one. Allow for bypassing the cleaner when cleaning is not necessary. Make BCFM (broken corn and foreign material) or dockage (small grains and fines) tests on incoming grain to help evaluate harvesting and variety performance.

### Damage from Drying

Desired grain quality limits drying air temperature. Because many factors are involved, only broad guidelines can be given for recommended drying air temperatures. When planning, consider what type of system and what drying equipment will provide enough flexibility to satisfy a range of drying needs. High-temperature grain drying (especially corn) followed by rapid cooling produces stress cracks and brittleness. Stress cracks are fractures inside the kernel that typically do not come through the outer seed coat. Cracking is greatest for corn when it is rapidly dried through the critical 19 to 14% moisture range and then is quickly cooled. Stress cracks reduce milling quality and increase kernel breakage during handling. Slow drying and delayed cooling reduce stress cracking.

*Other potential problems caused by high drying temperatures are*

- Reduced starch extraction for ethanol and wet milling or corn.
- Reduced baking and milling quality of wheat.
- Reduced germination of malting barley.
- Discoloration (browning) of kernels (heat damage).
- Fractured soybeans, when relative humidity of the drying air is below 40%.
- Reduced head yield of milled rice.

Physical grain quality is determined by maximum grain temperature reached, which depends on drying air temperature, airflow rate, drying rate, initial and final grain moisture content, and drying air distribution through the grain. Optimum drying air temperatures depend on dryer type, airflow rate, grain type, end use (feed, food, fuel, seed), and initial and desired final grain moisture content and quality.

For any high-temperature grain dryer, lower drying air temperatures maintain higher physical grain quality. Physical grain quality includes test weight, BCFM, color, stress cracks, fissures, and brittleness. Very high initial moisture content or drying to excessively low moisture content requires lower drying air temperatures to maintain grain quality. Poor initial grain quality can also require lower drying air temperatures. Generally, to maximize drying rate and fuel efficiency, dry at as high a temperature as you can and still be satisfied with the final grain quality. If quality is not satisfactory, reduce drying air temperature for any type of heated-air dryer.

### Drying Temperature Recommendations

Seed and food grains such as malting barley, edible beans, and rice need to be dried at lower air temperatures.

*Feed grains such as corn, sorghum and milo can be dried at higher air temperatures.*

- High-speed, automatic-batch, and continuous-flow dryers often operate at air temperatures 200 to 220°F for corn if the corn is cooled immediately in the dryer, and up to 240 to 250°F with delayed cooling.
- Continuous-flow and roof bin-batch dryers generally operate at 140 to 200°F.
- Heated-air on-floor batch bin dryers, with or without stirring, operate at 110 to 140°F on feed grains. If drying grain at depths exceeding 4 feet without recirculation or stirring, limit the temperature of drying air to 10°F above ambient air to prevent over drying.

*Consider the following procedures and concepts to help plan a system that does a good job of controlling a range of drying temperatures.*

- Dry seed grain at air temperatures below 110°F regardless of all other factors.
- Dry malting barley in high-airflow, automatic-batch, or continuous-flow dryers up to 120 to 130°F.
- Dry edible beans using drying air temperatures below 100°F to reduce splits and seed coat cracking.
- Dry food corn and commercial soybeans at temperatures up to 140°F in high-airflow column dryers, but not over 110°F in bin dryers. At these temperatures, stress cracks in corn are minimized, and splits and seed coat cracking in soybeans do not generally affect marketability. However, check the quality of food corn and soybeans frequently during drying, and reduce temperature if necessary.
- Dry small grains (wheat, durum, rye, etc.) for milling below 160°F in high-airflow batch- and continuous-flow dryers, and below 120°F in bin dryers.

## Damage from Handling

Increased grain breakage results from improperly installed or poorly operated handling equipment, grain impact at high velocities, and kernels stress-cracked and fissured during drying and/or cooling. Very dry or cold grain is more susceptible to breakage. Minimize damage with proper sizing and operation of handling equipment, low drop heights, and low grain velocities. Design the drying and cooling systems to minimize stress cracking and fissuring, making grain less susceptible to breakage during handling.

Operate screw conveyors (augers) at rated capacity to reduce grain damage, especially tube augers without internal bearings. With variable incoming flowrates, reducing the rpm of screw augers can keep conveyors operating at rated capacity. Choose bearing-supported conveyors where variable incoming flowrates make it difficult to operate conveyors at rated capacity at all times. Another option is to add a surge (accumulating) bin over the hopper to feed the conveyor, keeping it full.

Bucket elevators cause little grain damage unless there are large drop heights. When drop heights exceed 40 feet, grain damage significantly increases. Make sure spout angles are not too steep, and install grain decelerators in the downspout every 30 to 40 feet to slow the grain unless impact damage is of no concern. For seed grain and edible beans, bean ladders can be used to allow the material to flow instead of drop into the bin.

Pneumatic conveyors cause little damage unless improperly installed or operated. Avoid excessive grain velocities, rough or nonaligned tube joints, sharp turns, missing cyclones, and large drop heights. In planning your system, make sure there are 4 to 5 times the pipe diameter (inches) in feet of length between turns to maintain quality and capacity. For example, a 5-inch pipe needs at least 20 feet between turns.

## Damage during Storage

While planning to control grain quality by limiting damage from drying and handling, recognize the need to limit damage during storage. Most damage during storage is caused by molds and insects. Molds and insects grow best in damaged, improperly dried, or improperly managed grain. Plan to use aeration to control grain temperature. Pockets of fines (broken kernels, weed seeds, trash, and other materials) can cause spoilage because they restrict airflow and provide food for insects and molds. Clean grain to remove fines, and adjust grain spreaders to reduce concentrations of fines. Realize that broken kernels and fine material accumulate and concentrate under the downspout. Once a bin is peak-filled, ideally draw out the core of the grain mass until the surface is inverted by about one third of the diameter of the bin. This will substantially reduce the risk of grain spoilage due to self-heating of peaked grain and will assure uniform airflow through the grain mass. For more details, see the section on aeration in Chapter 5 of this publication.

CHAPTER 2

# LOCATING AND DEVELOPING A GRAIN CENTER

*Kenneth Hellevang, Ph.D., P.E., North Dakota State University*

The grain drying, handling, and storage facility is an integral part of harvesting. Its function is to effectively condition and safely store harvested crops for future marketing and should not constrain the harvesting rate of the producer. When locating and developing a grain center, it is important to recognize where some of the bottlenecks can occur during harvest and plan to minimize them. It is recommended and prudent to visit several existing grain centers to gather ideas and input on the pros and cons of system development from other operators. Suggested layouts also can be gathered from equipment dealers and contractors who have extensive experience with developing grain centers.

## EXPANSION OR NEW

Existing facilities need to be carefully examined relative to present and future needs. It is important to not allow existing facilities to influence the decision on how to build a new grain receiving, conditioning, and storage system. Carefully examine existing facilities when expanding or building a new system. An existing grain center may have limitations such as insufficient space for current and future expansion, and there may be other factors that help to determine whether it is feasible to incorporate existing facilities into a new, efficient system. Other limitations might include the capacity and condition of the existing material-handling equipment. For example, under-sized dryers, dryers without desired features, and bin size and location may limit efficient, economical use of existing facilities. Figure 2-1 shows a scattered layout typical of old grain systems. The drawing illustrates how smaller bins were located near the receiving hub when the grain system was first built. As the operation grew, larger bins were added but had to be located farther from the bucket elevator. Figure 2-2 illustrates a system with bins of various sizes using a portable-auger grain-handling system.

Consider building a new grain center when the existing facility cannot functionally support the harvest operation. A producer should consider a new facility at a different site when the existing facility cannot obtain adequate utilities (such as electrical and gas service) or does not permit safe, efficient grain handling. Building near existing facilities permits taking advantage of existing bins, but make sure existing bins fit with an efficient grain-handling system and can be an effective part of the new grain system. Consider using the existing facility as a satellite grain center, or convert it to handle an alternate grain, to a feed processing center, or to a long-term storage facility.

**FIGURE 2-1. A scattered layout in an existing system that may not fit with expansion.**

**FIGURE 2-2. Bins of various sizes loaded and unloaded with a portable-auger grain-handling system.**

## CENTRAL OR SCATTERED SITES

Grain facilities are usually at one central site, but there are many scenarios where it is appropriate to have remote and or scattered sites. Consider your farm's needs when planning the location of the grain center.

*Advantages of central facilities include:*

- More efficient grain handling for drying, storage, and management.
- Centrally located controls.
- Single electrical and fuel services for the entire system.
- Centrally located records and instruments.
- Less need to move equipment between sites.
- Centrally located security, including biosecurity.
- Fewer all-weather roads required.
- Reduced operational labor costs.

*Advantages of separated facilities include:*

- Smaller facilities that are easier to divide if the farm business breaks up or changes status.
- Less competition for expansion space with other farm enterprises.
- Shorter transportation distances between the field and facility.
- Decreased traffic around the main farmstead.
- Improved ability to segregate commodities.

More than one grain center may be needed if you operate on farms separated by long distances or if you have landlords who insist that their grain be stored on their land.

Consider a grain center near a central location determined to be central by travel distance adjusted by the volume of grain handled for the total farming operation, especially if fields are separated by several miles. Minimizing the distance to all fields saves fuel and hauling time and can eliminate bottlenecks during harvest.

When selecting a site, consider the resale value of the grain-handling site. If it becomes necessary to sell the site due to financial hardship, a change in family ownership, or other life changes, it is preferable to have the bin located away from the homes of family members.

## SITE SELECTION

*As you examine each probable site, ask questions like the following:*

- Is there enough space for future as well as immediate needs?
- What is the water table?
- Does the site have good drainage?
- Will soil type affect the facility?
- Is it accessible by all-weather roads that can handle large trucks?
- Will traffic affect neighbors?
- Can trucks safely and easily enter and exit the facility from public roads?
- What is the effect of system dust and noise?
- Is adequate and the proper type (single or three-phase) of electric power available, or can power be extended economically to the site?
- Is an adequate fuel (propane or natural gas) supply available for the dryer, or can a gas line be extended economically to this site?
- Will the site become a safety hazard for family members?

### Space Requirements

*The site must provide suitable space for the following components and activities:*

- Grain handling and cleaning equipment.
- Unloading and loading equipment (receiving pits, overhead loadout).
- Wet-holding, drying, cooling, surge, and storage bins.
- A horizontal, stacked, or tower dryer.
- Feed processing and storage.
- Fertilizer storage in coated bins.
- Electrical service and control.
- A management building for shelter, record keeping, and testing grain quality.
- A scale for weighing.
- Large trucks/semi-trailers moving into and within the facility.
- Fuel storage.
- Roads.

Not all these facilities may be needed at first, but consider providing adequate space in the plan for expansion and improvements. Later, you may need to include space to accommodate increased storage, a bucket elevator or other types of grain conveyors, wet-grain receiving, a new drying method, a switch to livestock or cash grain, a change in marketing plans, or feed processing.

Calculate the space needed for the planned grain bins, receiving area, dryer, conveying system, and roads. Plan space for future expansion (two to four times the original capacity). For example, could the site be expanded by clearing a nearby wooded area, removing a building, or converting some pasture or cropland.

Storage bins require a minimum separation distance (primarily dictated by the foundation footer), so contact a local authorized bin dealer to discuss these distances during the design of the facility. All mechanical systems require maintenance, repair and eventual replacement, so consider accessibility to mechanical systems during the planning phase.

## Drainage

Avoid areas with surface water runoff problems and high water tables. Select a site with surface drainage in all directions. If necessary, add earth fill to increase grade level. If surface drainage in all directions is not possible, divert water away from the grain center with waterways and terraces.

Regardless of the grain center's initial capacity or complexity, locate it and establish the grade based on future expansion. Move earth into an area well before construction begins.

## Soil

Verify the load-bearing capacity of the soil before the start of construction. Hire a local geotechnical engineer, or verify that the contractor has tested the soil and that soil information is included in the design. Also, determine if a geotechnical fabric will be required and what amount of fill and types of fill will be required.

## Snow and Ice

Evaluate the snow drop from the grain facility, shelter belts, and other buildings. How will the snow and ice be dealt with to permit using the facility during the winter? Consider the effect of snow and ice on grain conveying equipment. Also, consider the effect of melting snow on access to the facility and the need for drainage. Snow melt may affect where snow is piled during the winter. A southern exposure permits the sun to warm surfaces and assist with melting snow and ice during the cold seasons.

If possible, locate roads and unloading conveyors on the south side of bins and buildings to enable warming and melting by exposure to the sun.

## Wind

The facility will be used extensively during harvest, so primarily consider the prevailing winds that occur during the harvest season. Determine the predominate wind directions and the impact wind will have on the facility, especially at harvest. Where will grain dust accumulate and/or travel, both on- and off-site? Will blown dust affect your farm house or any neighbors? Will the wind affect grain unloading or loading? Will it affect dryer performance and the potential for fires? Position the dryer opposite to the prevailing wind to limit the wind's effect on the dryer's performance.

## Noise

Because fans and other equipment generate noise, create a sound barrier between the grain center and the house and other people. Select fans that produce less noise, and, if possible, position them so the sound travels away from houses, neighbors, and other areas occupied by people. Axial fans generally are the loudest, and low-speed centrifugal fans are the quietest. If possible, locate the fans and equipment away from any residence, and create a sound barrier, such as trees or other structures, between the facility and residence.

## Traffic

There will be lots of large trucks driving into and around the grain center site. The traffic will be greatest during harvest but will continue as the grain is marketed. A location remote from the house reduces traffic nuisance and safety problems in the living area.

Provide convenient access to the main road. Consider the weight of grain transport vehicles when locating the grain center to avoid light-load bridges and culverts. Consider a by-pass road to take non-grain traffic such as employees, family, or visitors around the grain center.

Avoid access routes prone to snow drifts or ponding water.

## Roads

Provide all-weather roads. Plan adequate space for snow removal equipment and snow piling.

Provide smooth traffic flow through and around all buildings and lots. Allow for expansion when locating and arranging roads. Farmstead traffic may not seem important, but when grain volume increases, transport vehicles get larger, loading time increases, and traffic delays can become a problem.

Plan straight access roads from the main road to the receiving/loadout area for easier maneuvering of loaded trucks. Plan all turns and loops properly. See Figure 2-3a for minimum road curve dimensions. Consider fencing and gating straight access roads that cannot be screened from public view (Figure 2-3b).

If the grain center is close to the road or in the corner of the intersection of two roads, consider more than one access road.

In your layout, plan roads so all grain-containing vehicles drive through the receiving and loadout area. Backing up is inconvenient, unsafe, and nearly impossible for high-volume grain handling. If possible, locate the receiving/loadout area so trucks drive straight when loaded and turn corners when unloaded. This will reduce necessary maintenance to the roadway. Locate the site components so that the personnel driving trucks exit the truck and step directly toward a driveway building or other commonly used work area. Not crossing a traffic pattern on foot will enhance safety, and exiting from the truck directly to the driveway building will add convenience and efficiency.

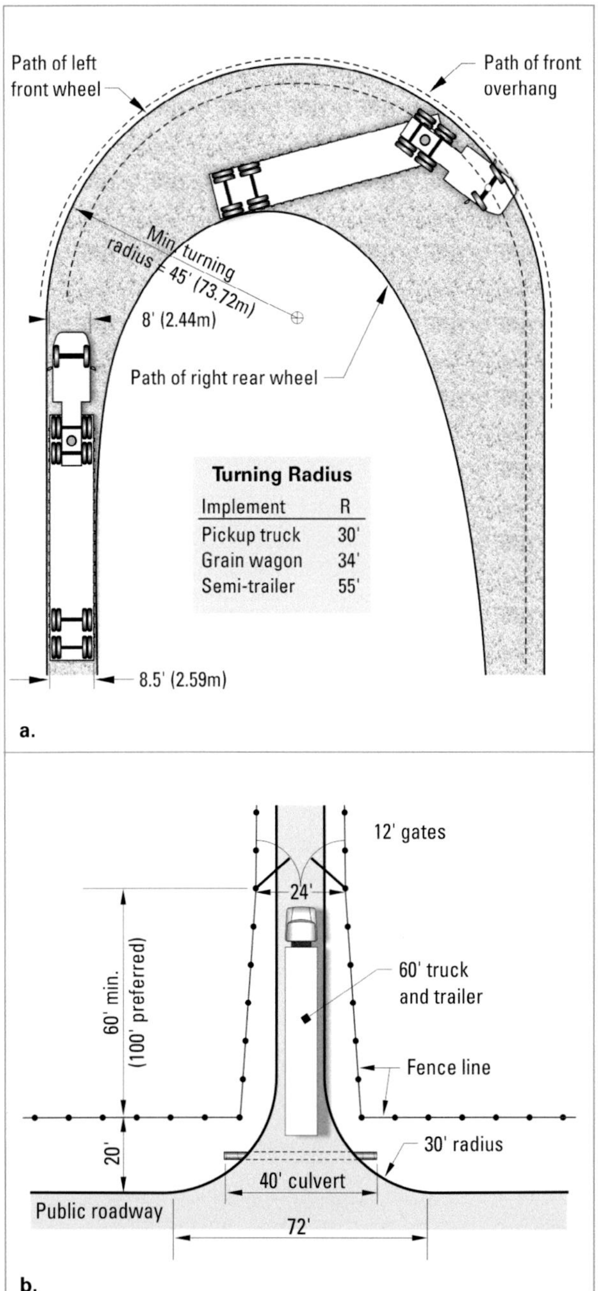


**FIGURE 2-3. Road curve dimensions.**

## Security

Reduce the risk of vandalism and theft with safeguards such as locked gates, locked unloading auger motors, lighting, motion-activated cameras, and other trespasser warning devices.

## Lighting

Loading and loadout often are carried into the night during harvest, and effective lighting not only makes the job easier but also improves safety and security at the site.

## Electrical Service

Electrical service is a significant factor in site selection. Contact an electrician and power supplier to discuss the availability and total costs for electrical service at each location being considered. Consider the availability of three-phase power and its impact on the facility. If future conversion to three phase is not feasible, consider phase converters, variable frequency drives, or soft-start motors for loads larger than about 10 horsepower. If three-phase service will soon be available, use phase converters and three-phase motors so the system can be easily converted. For convenience and safety, use underground wiring to and within the grain center.

Locate the grain center service entrance and control panel near the receiving, drying, and loadout facilities. Avoid the high humidity discharge air from the dryer. Place the panel and controls in a building when possible because weatherproof switches are more expensive than non-weatherproof switches. Consider locating them in the operations center building.

The control center should be centrally located so a facility operator can easily and quickly monitor the system. Monitoring grain flow at receiving and loadout is important. Consider installing pressure and level sensors, warning lights, and/or horns to alert the operator of grain flow or storage capacity issues. At a minimum, easy access to receiving and loadout areas needs to be considered for visual inspection.

The electrical requirements of a grain center are large and can be complex. Plan for all equipment in the final system even though some tractor-powered equipment may be used initially. Discuss with your power supplier available rebates for more energy efficient motors and dryers.

For more information about materials and methods for selecting electrical equipment and wiring in agricultural buildings refer to the publication MWPS-28, *Wiring Handbook for Rural Facilities*.

## Cleaning Equipment

*Consider arrangements for the following grain cleaning possibilities:*

- Wet grain into the system.
- Grain from dryer to cooling bin or storage.
- Grain from cooling bins to storage.
- Grain from storage to loadout.

In a well-designed system, one grain cleaner system can be used for several of these possibilities. Each type of cleaner performs a different function, so the desired action needs to be included in the design. The amount of material removed will depend on the type of cleaner and amount of foreign material in the grain. To help select the cleaner, obtain documented performance data for the cleaner being considered. A plan for how the foreign material will be collected and stored also needs to be included. Wet or warm screenings have a very short storage life, so they must be consumed or dried and cooled before they deteriorate.

## Bins

On-farm bins typically fall into one of six categories depending on function and structure. When siting bins, allow for expanding storage in at least one and preferably two directions, as illustrated in Figure 2-4. Bins can be arranged in one or two rows to be filled with horizontal conveying equipment from the central grain handling facility.

**Wet-holding bins** are usually filled directly from the receiving area and unloaded into a dryer. They are usually near the dryer and are sometimes positioned above it for gravity filling. Typically, wet grain must be ventilated to maintain grain quality until the grain has been dried.

Consider using a flat-bottom bin to increase wet-holding capacity in conjunction with an ancillary hopper-bottom, wet-holding bin. Increasing wet-holding capacity could reduce the necessary size of the dryer or shorten the workday of personnel.

**Surge bins** are used to collect grain and hold it for a short time. An example is an elevated hopper bin that holds the grain conveyed from storage bins at a slower conveying rate, and is unloaded rapidly to fill a large truck.

**Working bins** are bins, such as a surge bins, that are loaded and unloaded many times per year.

**Drying bins** require a limited grain depth and need equipment such as fans, heaters, and stirring devices, all depending on whether the bin is being used for natural-air, low-temperature, or high-temperature drying.

**Storage bins** will be filled once or just a few times each year. All storage bins require aeration. Storage bins typically have a flat bottom, but can have hopper bottoms. Larger capacity storage bins will have wall stiffeners to add strength to the walls.

**Flat storage buildings** typically are more difficult to fill and unload. They require more ducts to properly aerate the grain and are appropriate only for dry grain. They commonly are part of overflow storage unless permanent grain conveying equipment that integrates the building into the grain center is installed for loading and unloading.

Allow for overflow storage in your plan. Increasing yields, changes in marketing strategy, and/or government programs may require additional storage. In the plan, include space and equipment for loading and unloading overflow storage and a method to provide aeration for storage management. Consider a site near the grain center for easy filling. Overflow storage should be needed only periodically, so hauling grain to market may allow better use of a structure at times when overflow storage is unnecessary. Consider expansion needs when locating overflow storage.

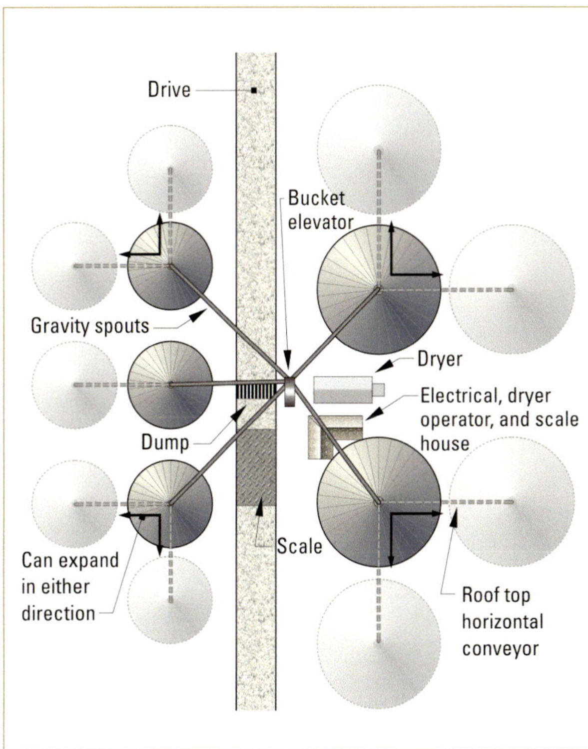


**FIGURE 2-4. Storage expansion.**
The dashed lines show possible expansion routes.

## Dryers

Dryers are typically near the system hub for easier conveying. Layout depends on the drying method, so consider present and future needs. As your operation expands, the drying method or capacity may change. See Chapter 4, Grain Drying, for a discussion of different drying methods.

Changing the drying method is more difficult than expansion. For example, a bin dryer needs different space, arrangement, and grain handling than a column dryer does. Consider possible locations, positioning, and possibly altering the handling system with a change in drying method.

## Fuel Storage

Large fuel storage reduces interruptions in the grain drying process. A 750- to 1,000-bushel per hour dryer may require 75 to 200 gallons per hour of LP. Drying 160,000 bushels of corn can require as much as 32,000 gallons of LP gas. To account for these kinds of needs, planning for a sufficiently large fuel storage capacity permits taking advantage of price breaks that often are available during low use periods. Assure timely fuel delivery to prevent interrupting drying.

Put fuel storage near grain receiving and drying. Provide easy access for delivery trucks, and follow safety, fire code, and insurance requirements when placing fuel storage tanks.

## Weigh Scale

*Most farms do not have a scale. However, include space to add a scale for these tasks:*

- Weighing grain in and out.
- Determining yields and calibrating yield monitors.
- Keeping loadout trucks under weight limits.
- Monitoring custom drying operations.
- Determining shares under a share-rent agreement.

Consider locating a scale in the grain receiving area. Scales are often located in the driveway where the truck dumps through the scale. A scale in the receiving area eliminates moving vehicles between weighing and dumping and reduces walking between the scale and the operations center. The scale house or office can be used to keep weight, moisture content, and drying records together. A building over the weighing and receiving area protects the scale from weather, prolongs scale life, and eliminates the need for a separate scale house.

Plan to put the scale in line with and adjacent to the dump pit. As detailed in Figure 2-5a, separate the dump pit and the scale by at least 3 feet to prevent grain spillage into the pit. With a covered receiving area, place the scale pit at least 3 feet from the door to minimize problems from blowing snow and rain.

Locate the scale away from the receiving area if marketing activity, weighing livestock feed, or custom grain drying conflicts with receiving. With separate receiving and weighing areas, the scale can be placed at least one vehicle length from the dump pit to permit weighing one vehicle while dumping another, illustrated in Figure 2-5b. Another option is to locate the scale along the vehicle entry-exit route to allow traffic to flow both ways across the scale.

If there is no plan to install a permanent scale immediately, consider a portable, one-axle scale to check truck wheel weights for road restrictions.

## Management Building

A building near the grain center can house controls, grain testing and electronic monitoring equipment, and written or electronic records. The building should provide space for moisture-testing equipment, grading screens, a desk, file cabinets, communications equipment, a computer, a coffeepot, and a cot. The management building should be insulated and heated. The management building frequently contains electric controls and panels that are referred to as a Motor Control Center. This building negates the need for weatherproof electrical equipment, thereby greatly reducing the costs of installing electrical equipment and controls.

## Grain and Feed Center

Most grain and livestock farms will have the feed processing facility separate from the grain center, with the feed processing facility having one or two bins for storing grain. The location of the grain center relative to the location of the present or future feed center is determined by the delivery system—vehicle or conveyor. With vehicles (portable grinder-mixer, self-unloading wagon, truck), transport distance is not critical unless it is excessive. Conveyor delivery (mechanical or pneumatic) over long distances usually needs modifications such as the installation of intermediate power drives.

Even if you do not currently need feed processing, plan for it. Planning for feed processing increases flexibility and value to the next generation and to prospective buyers. Selecting, locating, and arranging feed processing equipment is complex and is not included in this handbook. However, consider feed processing, delivery, and storage while planning.

Feed processing on livestock farms often starts with a portable grinder-mixer, which is competitive for feed volumes up to 400 tons per year. Portable grinder-mixers can be moved between sites to collect grain and feed ingredients, but it is best to bring the grain to the mill. Include space for a grinder-mixer and tractor in the grain center. Overhead gravity-flow bins, a weigh hopper to pre-weigh feed ingredients, and a storage area for base mixes and feed additives also can be incorporated. Develop a feed processing area or building for convenient handling of bulk grain and feed, and sacked or boxed micro-ingredients. If the volume of feed processing increases, consider a stationary grinder/mixer that fits in the same space as the portable unit for an easy transition.

Feed processing in the grain center works well on smaller, mixed crop and livestock farms. If the grain or livestock operations are, or are expected to become, highly intensified, large-volume enterprises, separating the grain center and feed processing is recommended. Stationary feed processing can be near the grain center or at a satellite location.

Satellite facilities, such as the one illustrated in Figure 2-6, are best for large operations to reduce traffic and space conflicts. With satellite facilities, livestock and grain operations are separate, except for replenishing the feed center. Plan a three-week to three-month supply of feed ingredients at the feed processing center. Small storages provide for more rapid draw-down, easier estimates of grain use, and easier monitoring of grain quality. With smaller bins, less special filling equipment is needed, and aeration is less critical.

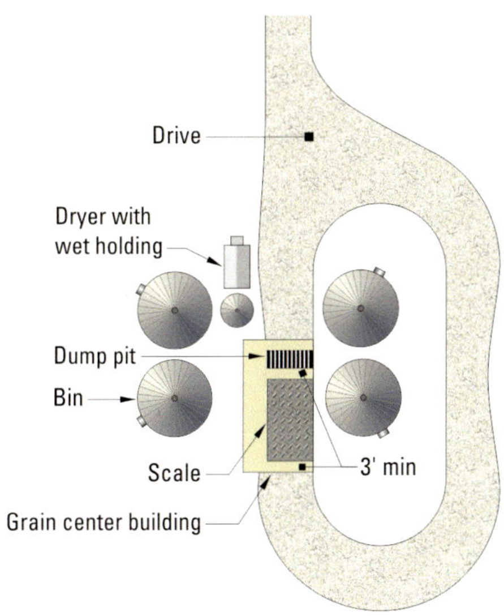


a. Scale at end of drive over dump pit.

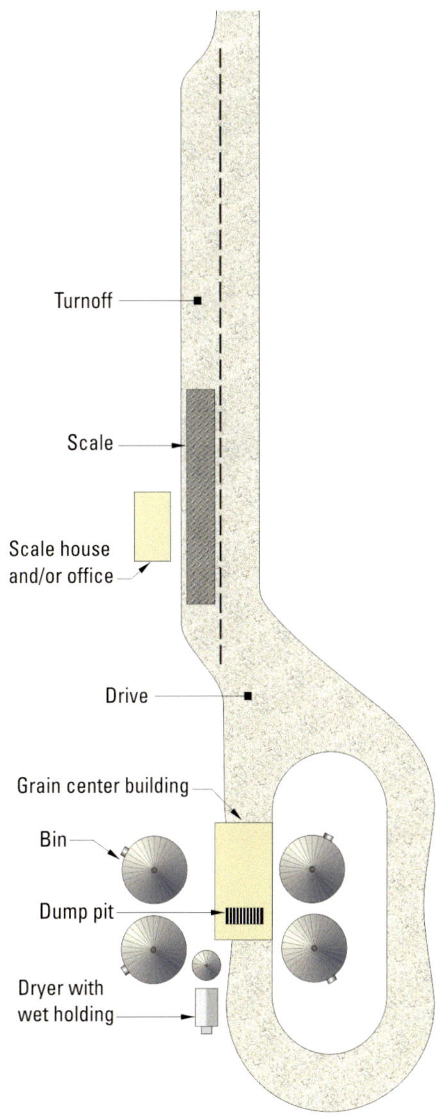


b. Scale located one truck length ahead of dump pit.

**FIGURE 2-5. Scale locations.**

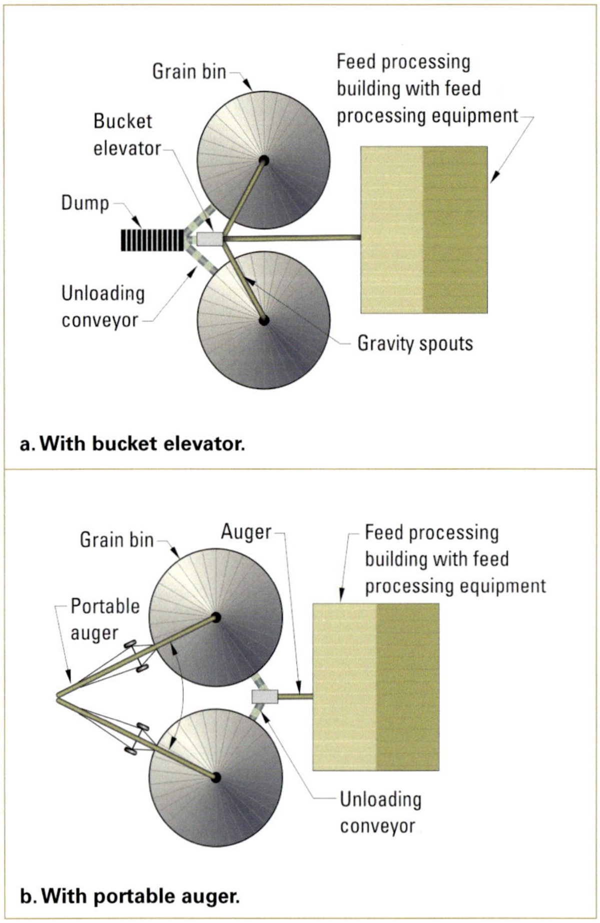


**a. With bucket elevator.**

**b. With portable auger.**

**FIGURE 2-6. Example of a feed-processing satellite facility, showing two possible conveyor options.**

# ARRANGEMENT AND LAYOUT

*In the layout plan, consider the arrangement of all the components:*

- Grain handling and cleaning equipment.
- Wet-holding, drying, cooling, and storage bins.
- A horizontal, stacked, or tower dryer.
- Fuel storage.
- A scale for weighing.
- A management building for shelter, record keeping, and testing grain quality.
- Feed processing and storage.
- Fertilizer storage in coated bins.
- Electrical service and control.
- Roads.

In your plan, always consider the need to remove screw auger flighting for a variety of reasons, including replacement, plugging, or inspection. Because auger flighting will need to be removed and replaced (pulled out from their tubes / U-troughs), allow one half of the bin diameter in front of the unloader. Also, arrange facilities to minimize walking, hand labor, and the need for brute strength. Arrange the system for easy visual and/or electric control monitoring. Most grain equipment operates automatically and continuously.

## Rough Sketching

Rough sketching is the best way to start developing a plan. Sheets of 8-1/2 inch by 11 inch plain or graph paper are easy to work with and are large enough to draw preliminary sketches for evaluation. Another option is to download 3D graphics programs to accomplish a similar task by computer.

The basic idea of rough sketching is to quickly draw a few alternative plans. Include future capacity needs and items such as a scale, grain cleaner, and overflow storage. The drawings should resemble actual equipment, but need not be detailed.

Do not spend a lot of time drawing facilities precisely. Measure only enough to be close. You will learn more from comparing four or five rapidly sketched alternatives than you will learn from one or two neat drawings that limit your options. Rough sketching is a trial and error process. It takes time and imagination. Rough sketching is a productive and relatively easy way to start, but it can also be frustrating. Many ideas may be sketched before one fits the guidelines of this handbook and the grain-handling and family needs for a specific farmstead.

Start by drawing an arrow indicating direction. North is usually at the top of the sheet. Draw in any existing facilities. Then, locate the future drive and show its direction.

Arrange facilities around the drive. Many grain centers start with a dryer and a few storage bins. Storages bins are usually in one or two rows along the drive. A bucket elevator or other conveyor is usually located beside the drive, and the dump pit is usually in the drive.

Another approach is to use cutouts of major items. Arrange them on a piece of paper with the future drive drawn on it. When you select a layout you like, draw around the cutouts, or place tracing paper over them and draw them.

Existing facilities can sometimes be incorporated into a plan being used in their existing role or another role. For example, a drying bin now can be a cooling or dryeration bin later, and storages that are auger-filled now can receive grain from a bucket elevator later. Self-contained dryers and wet-holding bins can be relocated as the grain center develops.

After completing the rough sketches, make a photo copy and add arrows to show grain flow from grain receiving to the dryer, from the dryer to storage, etc. Work toward grain flowing from the receiving/loadout hub to any component and back again.

Figure 2-7 is the legend for the symbols used in Figures 2-8 to 2-11, which show sketches that can be used to generate ideas for grain drying, handling, and storage systems. These are examples only that illustrate how different grain drying, handling, and storage methods can be organized into well-planned systems. They are patterns for planning; not all of these ideas will fit your farm or needs, but the basic ideas and principles can be applied.

## Scale Drawing

After deciding on the best one or two layouts, draw a top-view scale drawing of the equipment and buildings. Figures 2-8 to 2-11 show scale drawings (in addition to the rough sketches) that demonstrate the transition from sketch to scale drawing. The scale drawings in these figures show existing equipment and future equipment expected for each rough sketch.

Graph paper is recommended to make scale drawings. Use 11" by 17" sheets to include more detail. Use a circle template or compass to draw circular bins and a ruler or scale to draw the plan accurately. Pick a scale and record it on your drawing. For example, on graph paper, one square might equal 2 feet. If a ruler is used, different scales are available—1/16th, 1/8th, or 1/10th.

Since the sketches are not drawn according to scale, equipment on the rough sketches might not fit when drawn to scale. It may be necessary to relocate equipment several times before things fit together properly. Figures 2-12 through 2-15 were drawn to scale with computer software and show additional options to consider as the layout is being explored.

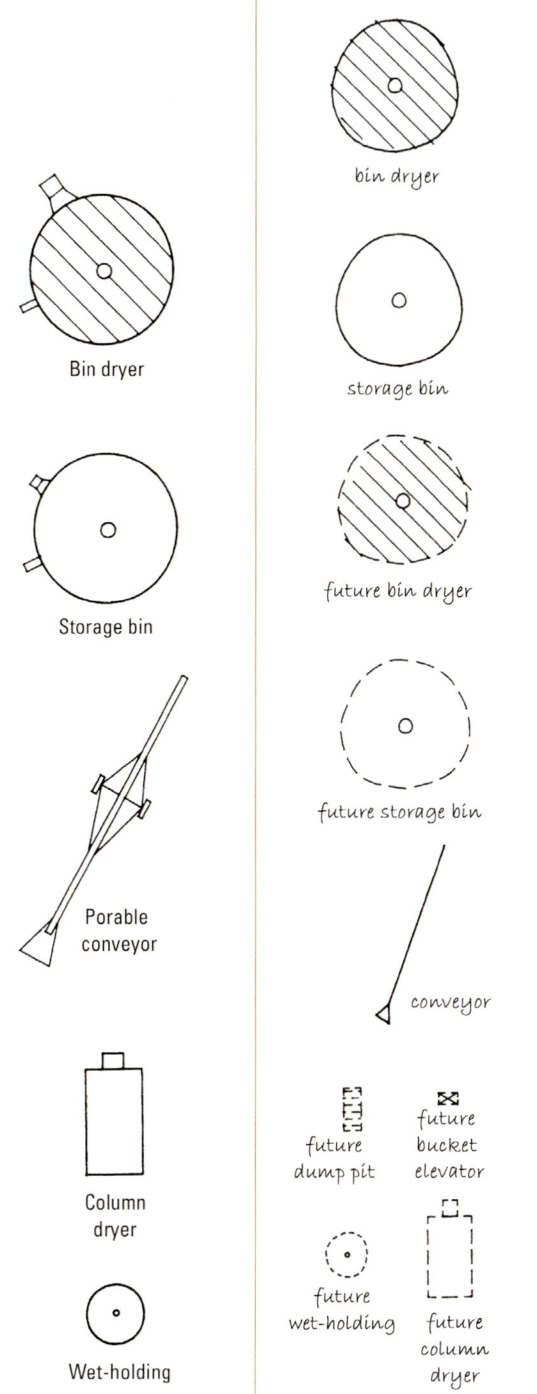


**a. Existing equipment** represented with solid lines.

**b. Future equipment** represented with dashed lines.

**FIGURE 2-7. Legend for Figures 2-8 to 2-11.**

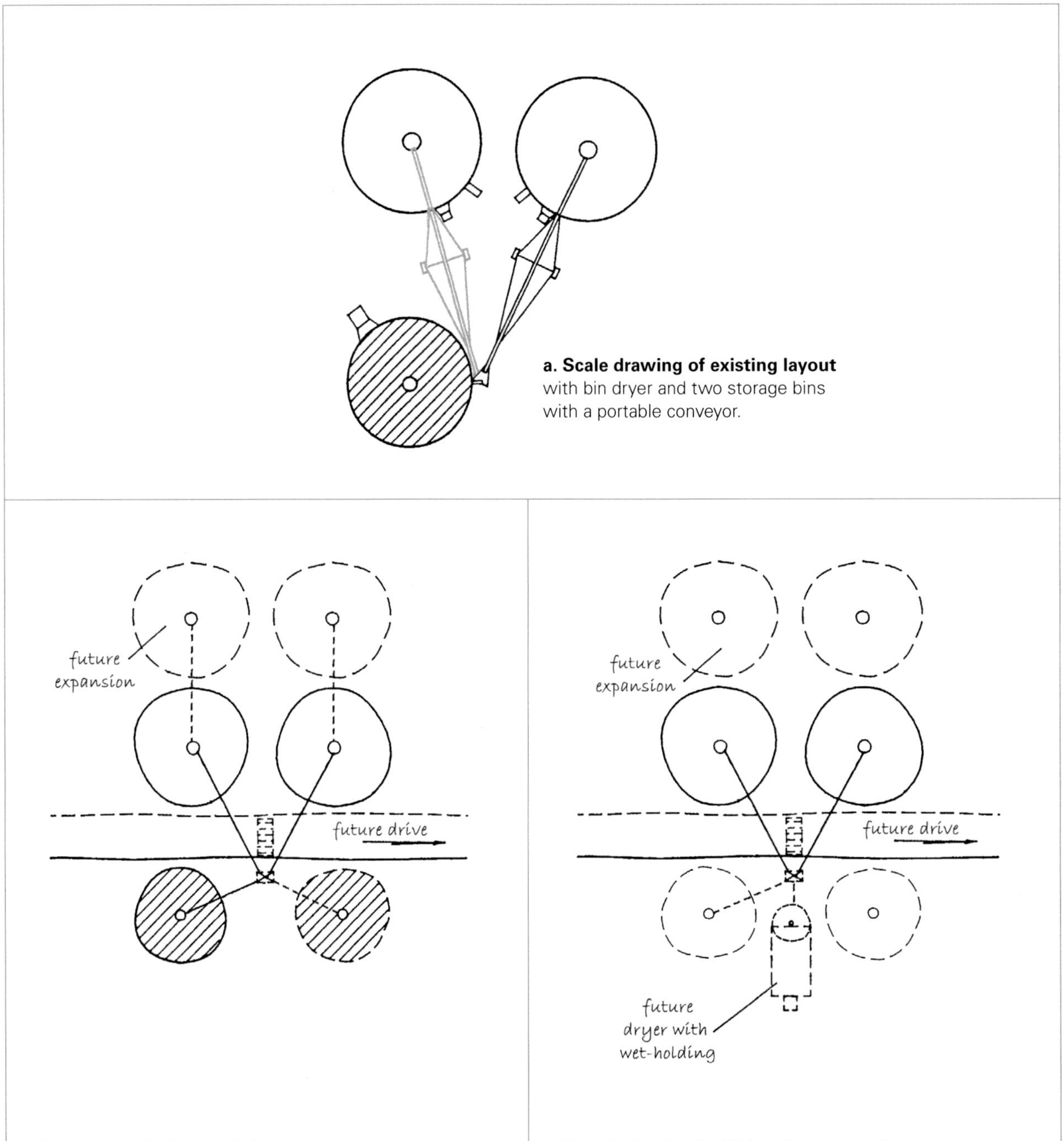


**a. Scale drawing of existing layout** with bin dryer and two storage bins with a portable conveyor.

**b. Rough sketch of potential layout** with two bin dryers, up to four storage bins, a bucket elevator, and a dump pit.

**c. Rough sketch of additional layout option** with a column dryer and wet-holding, and the two bins dryers converted to storage.

**FIGURE 2-8. Two future expansions from an existing bin dryer, to two storage bins and one portable conveyor (a), to a second bin dryer (b) and to a column dryer with wet-holding bin (c), additional bins, bucket elevator, and dump pit.**

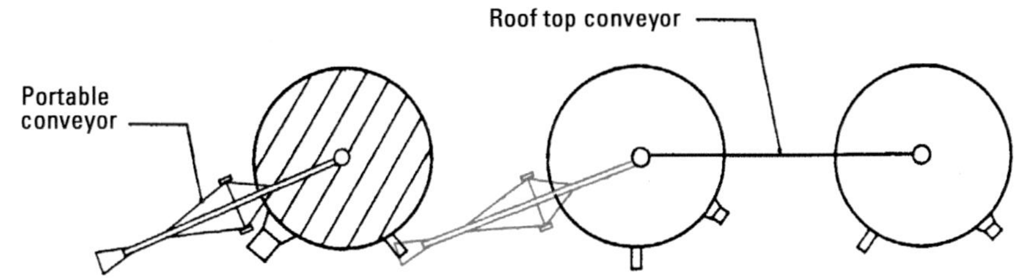


**a. Scale drawing of existing layout** with bin dryer and two storage bins with portable conveyor and roof top conveyor.

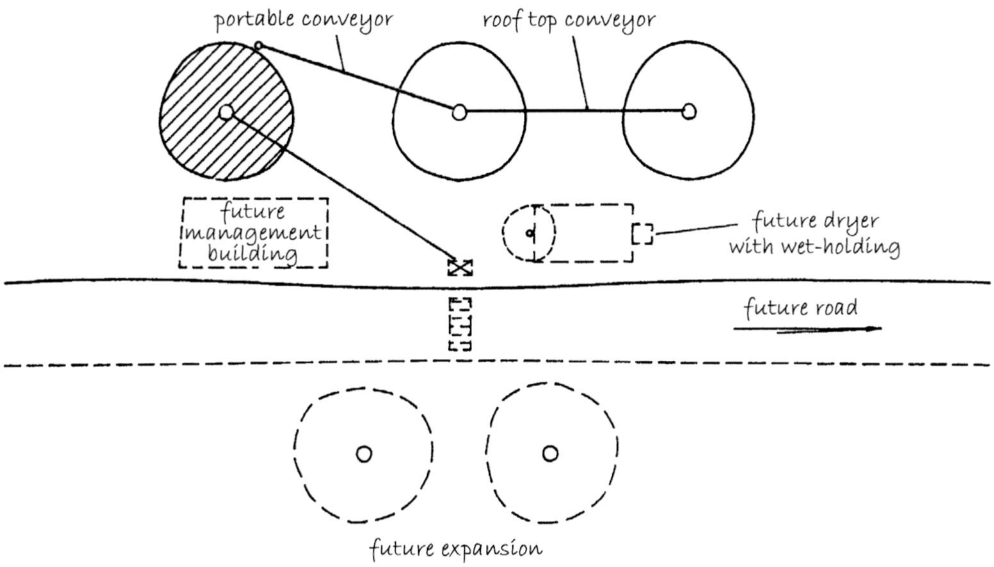


**b. Rough sketch of potential layout** with a column dryer and wet-holding, the bin dryer converted to storage resulting in up to five storage bins, a bucket elevator, a dump pit, a column dryer with wet-holding, and a management building.

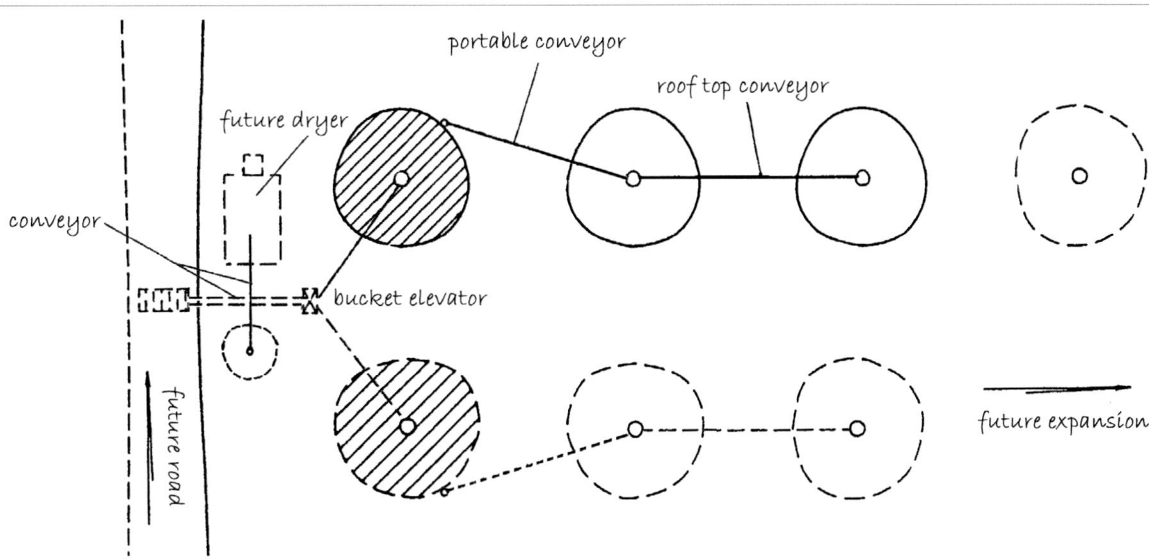


**c. Rough sketch of additional layout option** with same components resulting in seven storage bins or more.

**FIGURE 2-9. Two future expansions from an exising continuous-flow bin dryer, with two storage bins (a) to a system with a column dryer and five (b) and seven (c) storage bins, future dryer, and management building.**

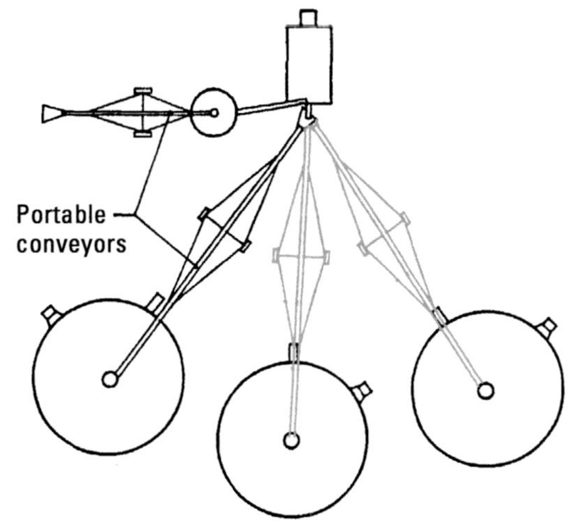


**a. Scale drawing of existing layout** with dryer and wet-holding bin, three storage bins, and two portable conveyors. The small wet-holding hopper bin on skids can be moved during future expansion and used elsewhere.

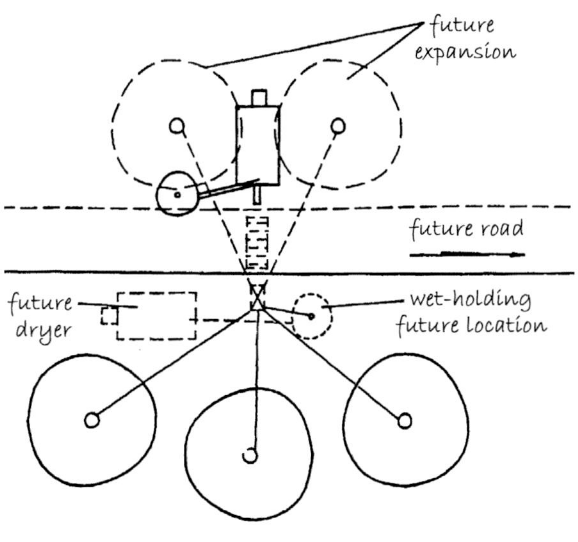


**b. Rough sketch of potential layout** with dryer and wet-holding bin, three storage bins, a bucket elevator, and a dump pit. A future option is to move the dryer and wet-holding bin to the other side of the drive and build more future storage bins in the freed up location.

**FIGURE 2-10. Column dryer with curved layout of bins.** Continuous-flow or batch dryers. Existing column dryer and portable auger can be moved when expanding to a central operations center with bucket elevator.

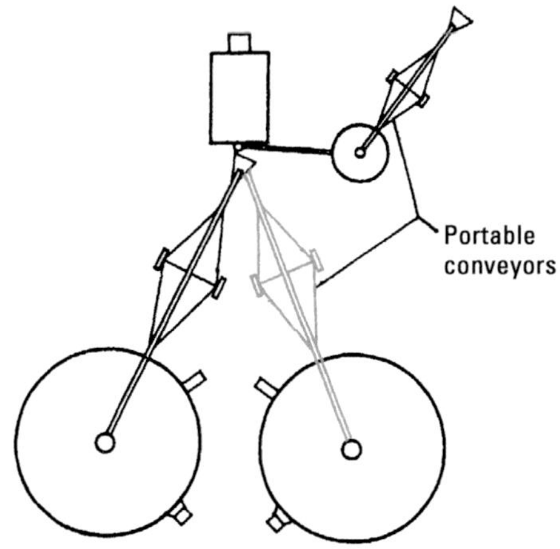


**a. Scale drawing of existing layout** with dryer and wet-holding bin, two storage bins, and two portable conveyors. The small, wet-holding hopper bin on skids can be removed during future expansion and used elsewhere.

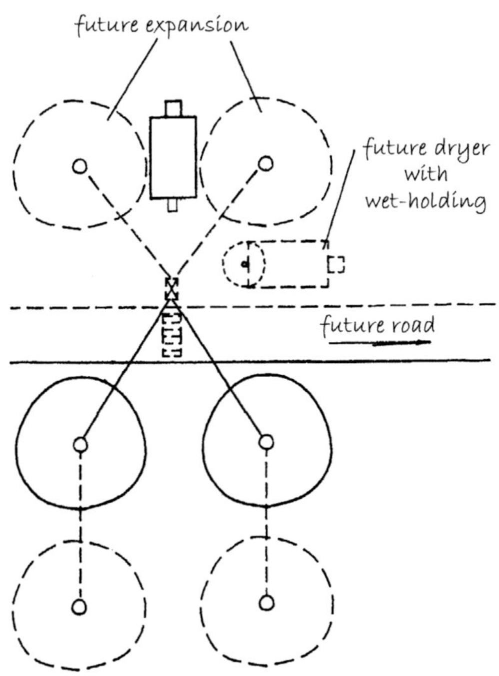


**b. Rough sketch of potential layout** with dryer and wet-holding bin, four storage bins, a bucket elevator, and a dump pit. A future option is to move the dryer with an overhead wet-holding bin sideways on the same side of the drive and build future storage bins in the freed up location.

**FIGURE 2-11. Column dryer with linear layout of bins.** Continuous-flow or batch dryers. Existing column dryer and portable auger can be moved when expanding to a central operations center with bucket elevator.

**FIGURE 2-13. Layout of grain center showing a chain-conveyor loop system to load and unload in-line storage bins.**

*Adapted from University of Missouri Extension "On-Farm Grain Storage and Drying Considerations"*

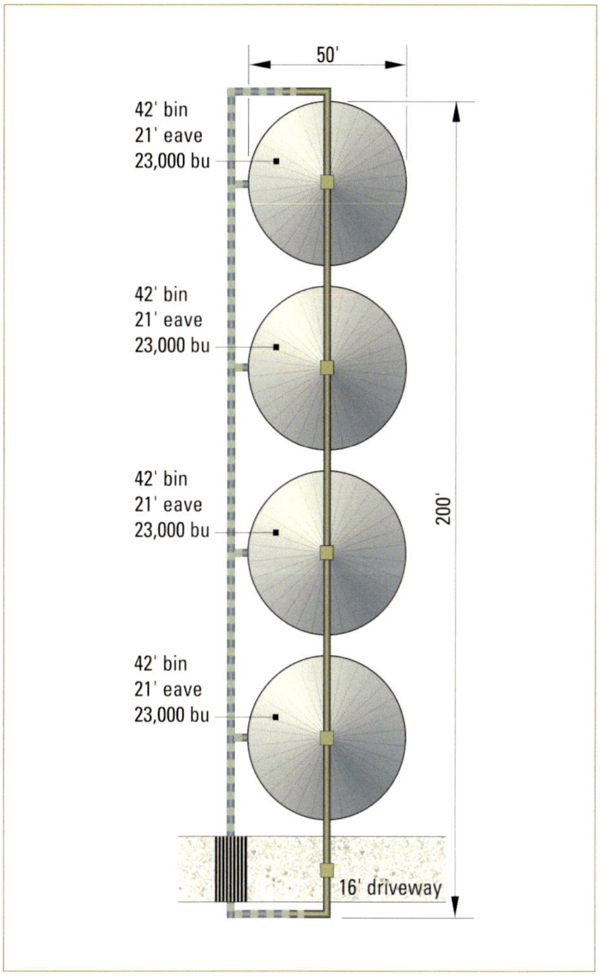


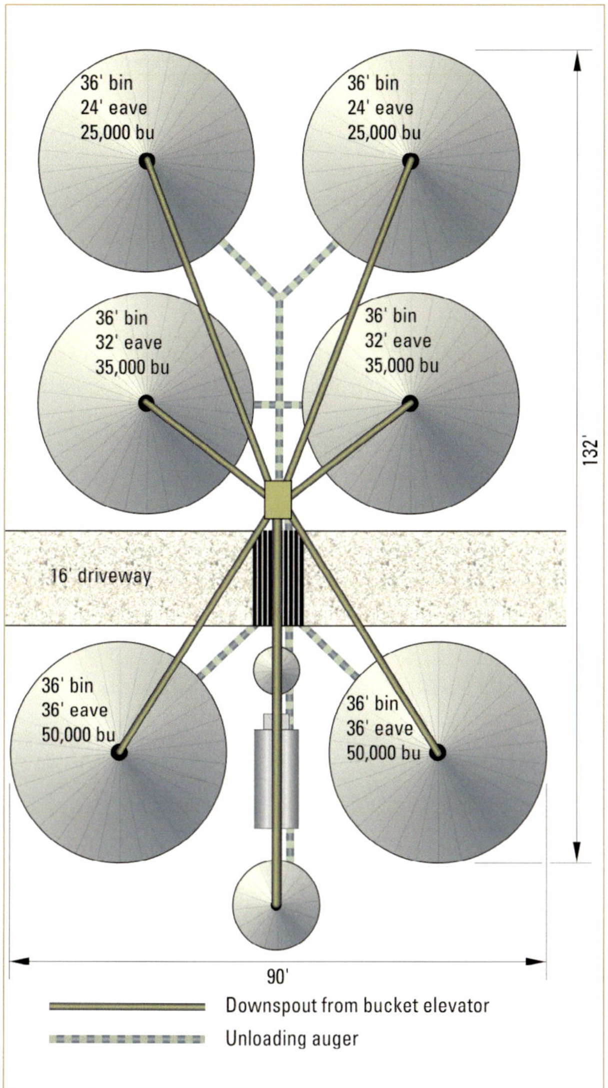


**FIGURE 2-12. Layout of grain center showing details of bucket elevator downspouts and bottom-unloading conveyors.**

*Adapted from University of Missouri Extension "On-Farm Grain Storage and Drying Considerations."*

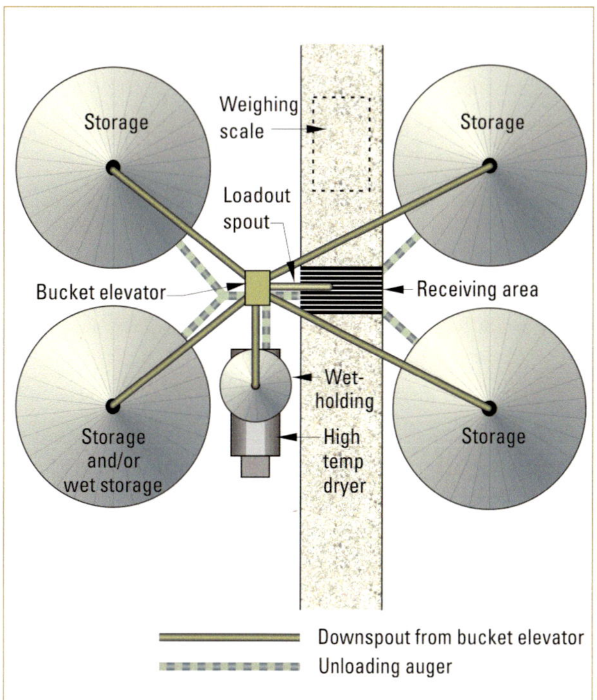


**FIGURE 2-14. Layout of grain center with a single bucket elevator and unloading augers to load and unload wet-holding storage bins and trucks.**

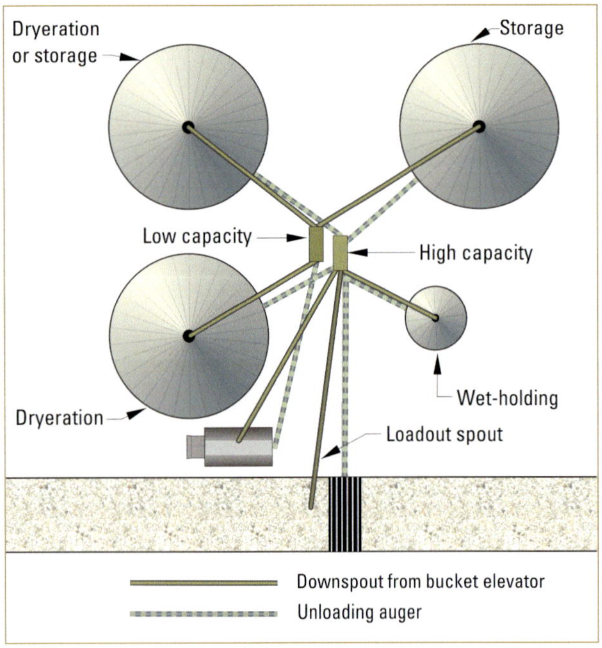


**FIGURE 2-15. Layout of grain center with two bucket elevators and unloading augers to load and unload wet-holding and storage bins, and trucks.**

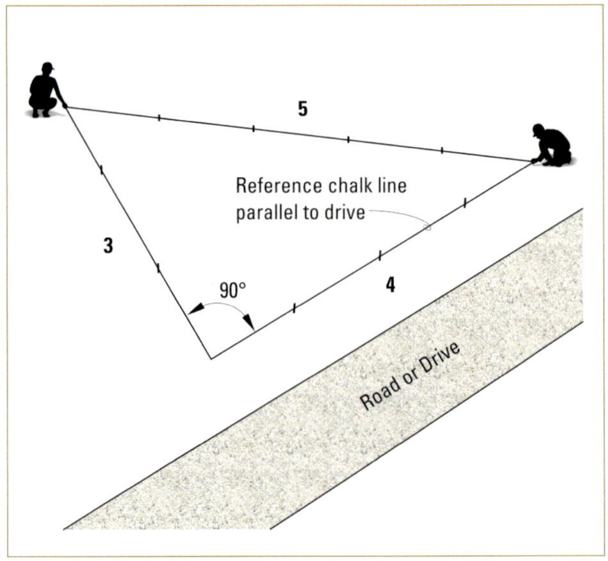


**FIGURE 2-16. Staking a 90-degree angle with a 3-4-5 triangle.**
A 90-degree angle is formed between the two shorter sides in any triangle where the side proportions are 3, 4, and 5. Example combinations are: 3'-4'-5'; 9'-12'-15'; or 30'-40'-50'.

## Staking

*Stake out the final plan on your site for the following reasons:*

- Staking the actual plan helps you visualize the layout better on the site than on paper. You may see that some changes in the arrangement are needed.
- Staking the actual layout helps ensure there are no mistakes. It shows whether all equipment fits as shown on the plan.
- Once staked out, you can drive through the traffic patterns. If possible, drive a semi-trailer through because it is difficult to maneuver in tight or improperly arranged traffic ways.
- Once staked out, it allows you to evaluate slope and drainage.

Two people can stake a layout with a 100-foot measuring tape, chalk line, and some stakes or flags. First, locate the future drive or other limiting boundary (road, tree line, or existing facility) as a reference for locating equipment. Use 3-4-5 triangles to form 90-degree angles to obtain lines that are perpendicular. Figure 2-16 illustrates this process.

Mark the outline of each piece of equipment with paint, lime, or other material. Do not forget to include future developments such as the expansion or modification of grain drying and storage, feed processing, grain cleaning, and scales.

## Evaluate the Plan

Evaluate your plan by going through the grain systems checklist, Table 2-1. All answers do not need to be excellent for an acceptable system, but do not accept any poor answers. Consider each question carefully. Do not make an error in judgment today that you may regret in the future. Use the checklist to identify both small and potentially serious problems. It is less costly and easier to correct a mistake during planning than during or after construction.

**TABLE 2-1. Grain systems checklist.**

| Excellent | Fair | Poor | |
|---|---|---|---|
| ❍ | ❍ | ❍ | The grain center site includes space for expansion and future improvements. |
| ❍ | ❍ | ❍ | There are no surface or subsurface water problems at the site. |
| ❍ | ❍ | ❍ | All-weather roads are planned for year-round access. |
| ❍ | ❍ | ❍ | Safety practices and considerations are recognized and will be implemented. |
| ❍ | ❍ | ❍ | A management building is planned to house electrical controls, grain testing equipment, etc. |
| ❍ | ❍ | ❍ | Additional storage will have good grain handling connections. |
| ❍ | ❍ | ❍ | Electric power requirements have been discussed with the power supplier. |
| ❍ | ❍ | ❍ | Grain center considers monitoring, safety, and theft protection. |
| ❍ | ❍ | ❍ | All equipment can be easily maintained and repaired. |
| ❍ | ❍ | ❍ | Fan noise and dust will not be a problem to the farm home or neighbors. |
| ❍ | ❍ | ❍ | Wet-grain receiving is fast and convenient. Grain moves conveniently between locations in the system. |
| ❍ | ❍ | ❍ | All bins and buildings are planned to be built on concrete foundations at least 12 to 16 inches above grade. |
| ❍ | ❍ | ❍ | Grain center is adequately spaced away from the farm home or neighbors. |
| ❍ | ❍ | ❍ | Adequate storage space and convenient access are planned for LP gas or other fuels. |
| ❍ | ❍ | ❍ | Roads and drives can handle or be adapted for semi-trailer trucks and commercial grain traffic. |
| ❍ | ❍ | ❍ | Drying methods such as dryeration, in-bin cooling, combination, natural-air, low temperature, and various types of self-contained high-temperature dryers can be easily adopted. |
| ❍ | ❍ | ❍ | Grain-cleaning equipment is included or can be added. |
| ❍ | ❍ | ❍ | A truck scale is or can be easily included in the grain center. |
| ❍ | ❍ | ❍ | Space is included or is available to build a feed processing, delivery and storage system. |
| ❍ | ❍ | ❍ | The grain center is appropriately located in relation to present or future livestock facilities. |
| ❍ | ❍ | ❍ | The grain center is located in reasonable proximity to fields to minimize travel at harvest. |
| ❍ | ❍ | ❍ | Equipment is oriented to take advantage of the winter sun and considers prevailing winds. |

## SELECTING EQUIPMENT SUPPLIERS AND CONTRACTORS

Initial cost is important, but selecting reliable builders and suppliers that properly install and promptly service equipment when required is even more important. A more expensive dryer, conveyor system, or other equipment that performs well is better than less expensive equipment that performs poorly or breaks down frequently. Find out how long equipment suppliers and contractors have been in business, and check with others to find out their experience with the equipment supplier or contractor. Generally, those in business for many years have done good work and promptly provided service. Local dealers and contractors often want to do a good job because they want future business from you and your neighbors.

Problems often result from poor communication. Verbal communication is easy to misinterpret. Always put the agreement in writing, and have it signed and dated. Too often the only signed and dated paper is the invoice of equipment to be supplied and the price. Other important points to be agreed on in writing are that the equipment will be installed according to the manufacturer's specifications and that it is guaranteed to operate as specified by the salesperson.

**Examples of important points of agreement:**

a. A dryer will dry a specified number of bushels of grain per hour removing a specified amount of moisture under specified ambient air conditions.
b. Conveyors will move a specified number of bushels of grain per hour. In addition, chain- or screw-conveyor speed should be specified. Consider the cost of various conveyor sizes relative to conveying speed, because the service life of flighting and bearings is related to chain and screw speed.
c. Each conveyor in a conveying series will move 5 to 10% more grain than the previous conveyor to minimize plugging and grain damage.
d. Drying, cooling, and aeration fans will deliver a specified airflow rate (cfm/bu) at a given static pressure.
e. Bucket elevators—the buckets' (cup) capacity should be calculated based on a level fill. Review capacity versus belt speed. Consider features such as access platforms or ladders/stairways. Will the bucket elevator be supported by cables or by a self-supporting tower? If by cables, consider their impact on road access and truck movement.

*Before construction starts, also secure agreement in writing to the following conditions:*

- All construction will be completed and equipment installed, in working condition, by a specified date.
- A payment procedure is specified in the contract.
- Disputes will be resolved by a mutually agreed upon arbitrator.
- Topsoil will be removed, and the soil will be prepared according to the equipment manufacturer's or an engineer's specification. Structures installed on farms today require special soil preparation and foundation design and construction to prevent uneven settling and even structural failure.
- Concrete will be placed on a properly prepared soil. Concrete will crack due to settling on an improperly prepared surface.
- A plastic vapor barrier will be placed under concrete, particularly under a bin floor. This practice reduces moisture problem in the bottom of bins.
- Unless otherwise agreed upon, concrete bin floors will be at least 12 inches above grade. Building floors and concrete pads under dryers, fans, and hopper bottom bins will be at least 6 inches above grade.
- Concrete will be reinforced with appropriately designed reinforcement such as rebar or welded wire to reduce cracks from uneven settling and to prevent cracks from opening wider.
- The mix, curing procedures, and finishes (troweled, broomed, floated, etc.) of all concrete floors will be agreed upon.

### Plans, Specifications, and Contracts

Detailed documents help provide needed communication, understanding, and agreement between owner, builder, and lender. Plans show all necessary dimensions and details for construction. Specifications support the plans; they describe the materials to be used, including size and quality, and they should outline procedures for construction and quality of workmanship. The agreement is a legally binding contract between the builder and the owner; it includes price of construction, schedule of payments, guarantees, responsibilities, and starting and completion dates.

*You have several options for preparing plans and specifications.*

- Be your own contractor. Draw a final plan including the specifications. Make sure dimensions are correct and construction details and materials are determined. Determine total costs before beginning

construction. Have plans checked by the appropriate regulatory agency when required. It is critical to have the plans evaluated by a professional to assure that all aspects of constructing the facility are done properly.
- Being your own contractor, including hiring all subcontractors and providing construction supervision, will be challenging unless you are experienced at managing construction projects.
- Use a design and construction firm. Some firms make working drawings and specifications and have standard contract forms.
- Hire a consulting engineer.

## Consulting Engineers

*Consulting engineers commonly offer these services:*

- Direct personal service (technical advice, etc.).
- Preliminary investigations, feasibility studies, and economic comparison of alternatives.
- Design.
- Cost estimates.
- Engineering appraisals.
- Bid letting.
- Construction supervision and inspection.

Consulting engineers usually do a project in three phases: preliminary planning, engineering design, and construction monitoring. They may be retained to help with one or more of these phases.

*To select a consulting engineer, consider the following:*

- Registration. To protect the public welfare, states certify and license engineers of proven competence. Practicing consulting engineers must be registered professional engineers in their state of residence and must be qualified to obtain registration in other states where they offer their services.
- Technical qualifications.
- Reputation with previous clients.
- Experience on similar projects.
- Availability for the project.

CHAPTER 3

# GRAIN **HANDLING**

*Kingsly Ambrose, Ph.D., Purdue University*

Well organized components make a grain drying, handling and storage system workable. Components include conveyors, hoppers, pits, surge bins, storage bins, spreaders, sweep augers, and cleaners. They receive grain, move it to and from any part of the grain center, and loadout the grain.

When selecting components, the most important factors to consider are performance, capacity, and convenience. Performance means getting the overall job done and not just having enough capacity to handle daily harvest and marketing, but also having the capacity to produce quality grain with a practical level of supervision. To accomplish these goals, match the overall capacity of the grain center to daily and seasonal farm needs. Size conveyors, hoppers, and other handling components to keep up with harvesting and drying. Other factors included in the selection of handling equipment include safety, reliability, cost, maintenance, and product damage. When thinking about acquiring and organizing grain handling components, a good principle to follow is to develop expansion potential into your facility design in the beginning; this will help minimize bottlenecks as the volume of your annual operation increases.

## CONVENIENT GRAIN HANDLING

Electrically powered, permanently installed conveyors provide the most convenient method of handling grain. They move grain at any time to any place despite weather conditions or other farm activities. Because no setup time is necessary, you can schedule low-capacity grain conveying needs as easily as high-capacity needs. With automatic control, conveyor capacities can be lower because continual supervision is not required.

All grain flow should be mechanized. You can start with portable augers to reduce costs, but plan for permanent conveyors as you get older, as original equipment wears out, and as grain volume increases. Good planning allows you to move gradually from low investment to a more ready-to-go, convenient system as profits and growth permit.

Develop the receiving/loadout area as the hub of the grain center. Plan equipment so grain can go from the hub to any component (for example, wet holding, dryer, storage) in the grain center and from any component back to the hub. Bucket elevators with downspouts and/or horizontal conveyors to bins are generally the hub for larger grain centers.

In a small center, grain may not need to return to the hub from every component. But as the system becomes larger and more complicated, return conveyors become more important. If return conveyors are not included in the original design, plan the system so they can be easily added later.

### Equipment Alternatives

Grain centers include wet-holding bins, high-temperature and high-capacity dryers, and/or drying bins, cooling bins, storage bins, cleaning equipment, and surge bins. Some processes are combined: a batch- or continuous-flow dryer can dry and cool; a drying bin with stirrers can dry, cool, and store. And, one process may utilize two units; combination drying starts in a high-temperature dryer and finishes in a low-temperature bin dryer. A component planned for one purpose may have alternative uses that do not seriously impair the original use; an elevated hopper-bottom bin can be used for both wet holding and dry-grain loadout.

Grain flow diagrams or sketches help prevent mistakes during early planning. Figure 3-1 shows common grain center functions and interconnections. The arrows represent conveyors. Table 3-1 is a list of equipment choices for the main functions illustrated in Figure 3-1.

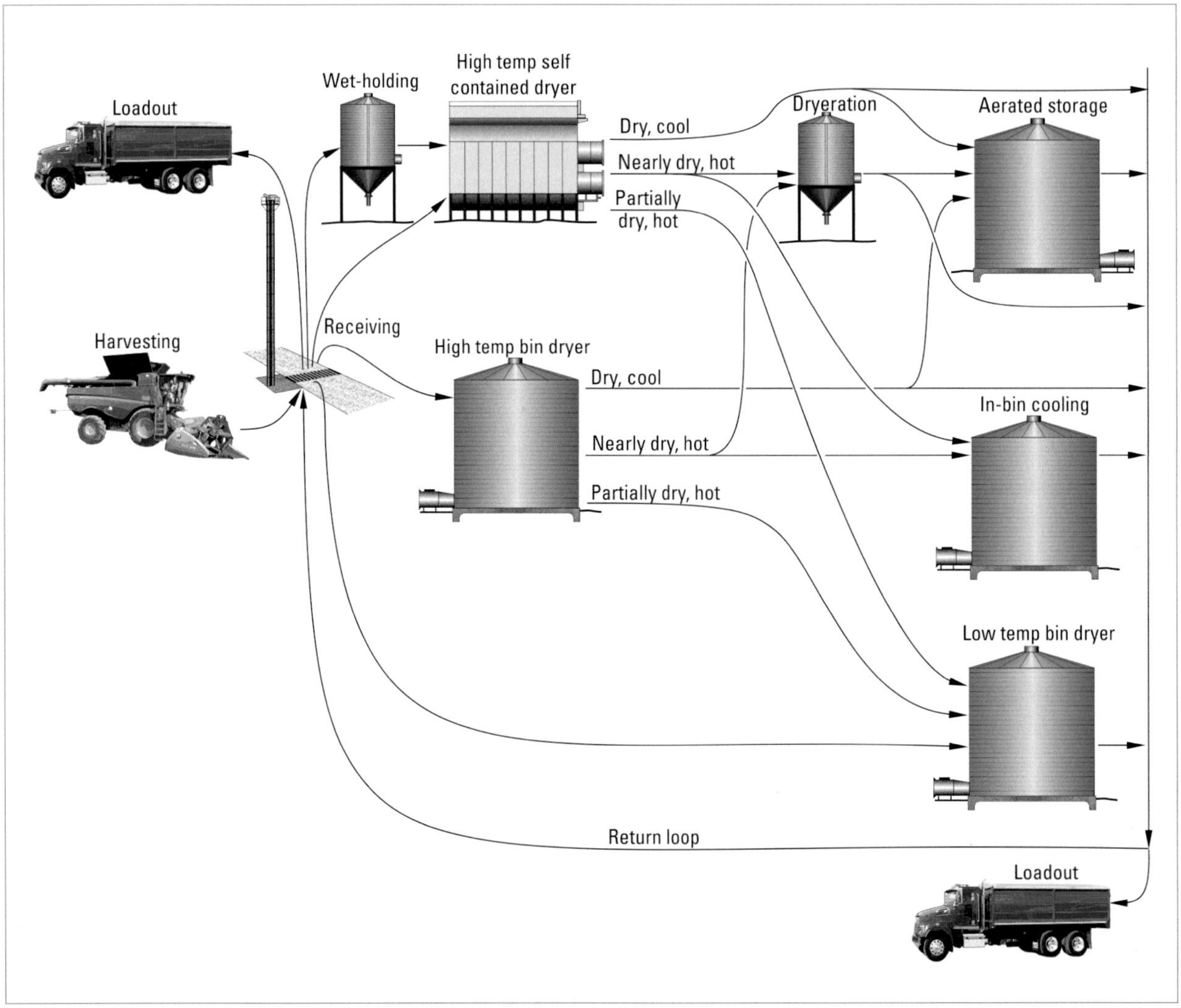


**FIGURE 3-1. Grain center components and conveying processes.**
Arrows represent conveyors.

**TABLE 3-1. Equipment alternatives.**

| Receiving | Wet-holding | Drying | Cooling after drying | Storage | Conveying |
|---|---|---|---|---|---|
| • Conveyor in a separate hopper<br>• Hopper attached to conveyor<br>• Portable, self-powered hopper<br>• Hinged or swinging conveyor attached to primary conveyor<br>• Drive-over conveyor<br>• Drive-over pit with conveyor | • Ground-level hopper-bottom bin<br>• Elevated hopper-bottom bin<br>• Flat-bottom grain bin with adequate airflow and fast unload/cleanout<br>• Gravity wagon<br>• Truck trailer | • Low-temperature bin dryer<br>• High-temperature bin-batch dryer<br>• High-temperature continuous-flow bin dryer<br>• High-temperature self-contained batch dryer<br>• High-temperature, self-contained continuous-flow dryer<br>• Combination high-temperature/ low-temperature dryers | • Cooling in dryer<br>• Storage bins equipped for cooling<br>• Bins or tanks equipped for dryeration and/or cooling | • Round metal bins<br>• Flat storage buildings<br>• Converted corn cribs<br>• Converted silos<br>• Temporary tarped bunkers—first grain to market<br>• Silo bags—cattle feeding or first grain to market<br>• Temporary tarped outdoor piles—first grain to market | • Screw augers<br>• Bucket elevators<br>• Flight elevators<br>• Pneumatic conveyors<br>• Bulk or mass flow drag conveyors<br>• Belt conveyors<br>• Gravity flow-fast loadout of market grain |

# GRAIN CONVEYING CAPACITY

Match conveyor capacity to present needs, but allow for increasing needs. Consider buying a larger capacity conveyor and running it full at a lower speed; speed it up later to increase capacity. Compared with a small, high-speed unit, a larger, slower conveyor lasts longer, has better power efficiency, and minimizes grain damage.

When one conveyor discharges into another, operate the second conveyor at higher capacity (10 to 15%) than the first (multiply by 1.10 to 1.15), so the second conveyor does not plug. For example, discharge a 1,000 bu/hr conveyor into one with 1,100 to 1,150 bu/hr capacity.

Electrical pressure or plug flow switches can be used to prevent plugging or overflow when one conveyor discharges into another. Install two switches in the second conveyor's hopper: a high-level switch to shut off the first conveyor when the hopper is full and a low-level switch to turn on the first conveyor when the hopper is almost empty. These switches are less expensive than cleaning spilled grain and unplugging conveyors. Motors overheat if cycled on and off frequently, so make the hopper large enough or the flow rate difference small enough so the motors run for at least 4 or 5 minutes per cycle.

You may need both high- and low-capacity conveying. Receiving and loadout require high-capacity conveying, but flow from a continuous-flow dryer is relatively low. Widely different rates require either different capacity conveyors or surge bins. Use electrical high- or low-level pressure switches to control flow into and out of surge bins.

*Surge bin uses include*

- Dump pits, so a truck or wagon unloads quickly and the pit empties over a longer period.
- A batch dryer unloads quickly into a surge bin, which delivers grain to storage at a slower rate.
- Overhead bins fill with low-capacity conveyors but empty quickly by gravity into a dryer or truck.
- A surge bin between a continuous-flow dryer and the dry-grain conveyor allows the conveyor to run intermittently and keeps the conveyor (particularly an auger) full, thus reducing grain damage. A surge bin can also accumulate dry grain while the conveyor moves wet grain.

Surge bins can be below-ground pits, gravity-flow wagon boxes, small storage bins, or commercial steel hopper-bottom bins. Hopper-bottom bins are the more costly alternative, but they can completely unload grain without mechanical equipment.

Gravity discharge is fast and convenient, but it is not free. Gravity unloading from a surge bin requires elevating grain higher than does direct delivery to the same elevation by conveyor as shown in Figure 3-2a and Figure 3-2b. Elevated bins are expensive, so use them where gravity flow has a big advantage and the bin is refilled many times a year.

Examples include wet-holding bins over dryers, bins over feed grinders, and bins to load trucks.

Commercial hopper-bottom bins are rarely justified for bins unloaded only once or twice a year. A bin sweep unloader in a flat-bottom bin does the same job without the space loss and high cost of the hopper-bottom bin.

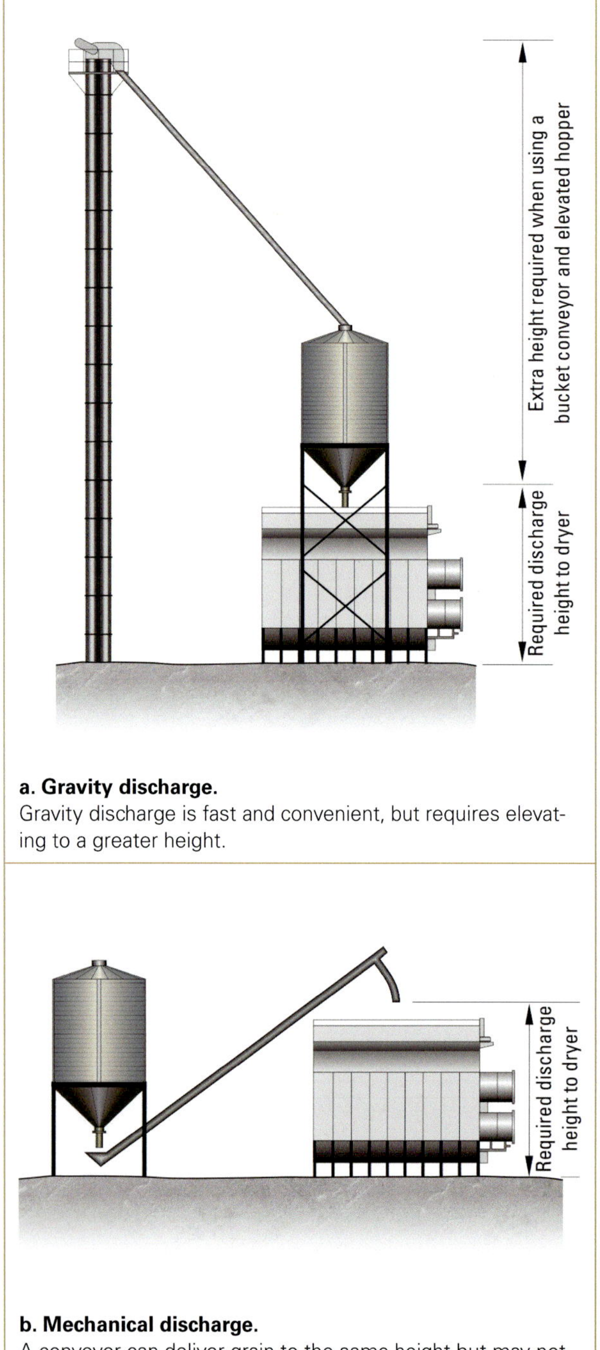


**a. Gravity discharge.**
Gravity discharge is fast and convenient, but requires elevating to a greater height.



**b. Mechanical discharge.**
A conveyor can deliver grain to the same height but may not be as convenient or as fast.

**FIGURE 3-2. Gravity vs. mechanical discharge from wet-grain hopper bin.**

# GRAIN HANDLING COMPONENTS

The process of grain handling requires equipment and space for receiving, conveying, spreading, cleaning, and loading grain, and for cleaning grain bins. Other issues to consider when evaluating grain handling components are the system's capacity to receive grain and how grain impact and grain velocity affect grain breakage.

## Grain Receiving

Harvested grain enters the grain center in the receiving area and is transferred to a wet-holding bin, dryer, or perhaps directly to storage. Grain receiving is often a bottleneck, so plan to provide high receiving capacity to speed up the unloading of trucks or wagons. Portable, above-ground receiving hoppers, such as those shown in Figures 3-3a, 3-3b, and 3-3c usually cost less than recessed or drive-over units. However, for each grain load, they must be moved or backed up to, and spilled grain must be scooped into the hopper.

Dump pits can be constructed of reinforced concrete or welded steel. The structural design of pits and grates varies with loads and dimensions; contact a knowledgeable engineer for structural design. Below-ground pits require special attention to drainage, or they must have a rain cover.

Grain is moved from the dump pit, shown in Figure 3-4, to the elevating inlet boot of the conveyor by a conveyor or by gravity. A small (less than 50-bushel) dump pit can be the hopper for a high-capacity auger, drag conveyor, or bucket elevator. Small dump pits are usually shallow to reduce groundwater problems. A dump pit as big as the largest transport vehicle permits fast unloading and can be a surge bin for a relatively low-capacity bucket elevator, auger, or pneumatic conveyor, as well as a wet-holding bin for a small dryer. Large dump pits can lead to deep excavation and possible groundwater problems.

A cover grate over dump pits is essential for safety, and a grate allows driving over the pit. The grate area must be large enough to receive grain without spillage. Backing a truck or a wagon to unload is dangerous and inconvenient, so plan vehicle access accordingly.

Dump pit augers can overload if too much flighting is exposed to grain; cover the exposed intake flighting as needed. Adjustable gates, baffles, half-pitch flighting, tapered diameter, or auger sections with smaller or tapered flighting also reduce the overload on pit augers.

Pave at least an 8-foot by 8-foot section of the grain receiving area with 6- to 8-inch thick, reinforced concrete. Divert surface runoff around the grain center. Slope the concrete at least 1/2 inch per foot away from the dump pit

**a. Bucket elevator with swing-away above-ground dump pit with screw auger.** Other options include hopper belt or drag conveyor to bucket elevator.

*Photo courtesy of L Cubed Corporation dba TAM Systems.*

**b. Portable inclined belt conveyor with fold-down above ground dump pit with belt conveyor.**

*Photo courtesy Ray-Man Inc.*

**c. Portable inclined screw auger with swing-away above-ground dump pit with screw auger.** Other options include belt or drag conveyor to screw auger.

*Photo courtesy of Farm King.*

**FIGURE 3-3. Above-ground dump facilities.**

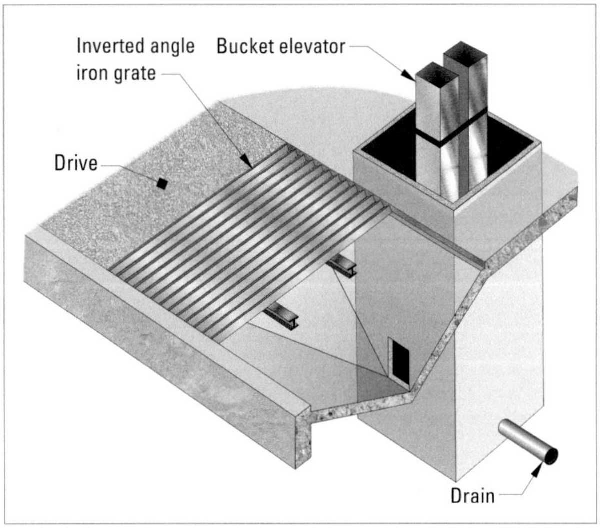


**FIGURE 3-4. Drive-over below-ground dump pit.**
Bucket elevator is gravity-fed from gravity pit.

opening. Keep rain and snow out of the pit with a solid cover. Provide drainage in the bottom of below-ground pits. Protect deep pits from groundwater by installing drain tile around the pit a few inches below the lowest point of the pit. Drain the tile by gravity or sump pump. If your water table is high, build a shallow dump pit on built-up earth fill for better drainage, as illustrated in Figure 3-5.

A building over the receiving area gives all-weather protection and convenience. This building can house management, control, and feed processing facilities. If necessary, leave space, and plan the layout to add a building over the receiving pit later. Provide at least 3 feet of space from the door to the edge of the pit to reduce rain and snow blow-in. Recommended doors are 14 feet by 14 feet, but should never be less than 12 feet wide. Make doors 16 feet wide for large tractors with dual wheels hauling gravity wagons.

A dump pit near one end of the building lets long trucks with hydraulic-lift beds park outside to dump grain. For indoor truck hoisting, run the roof trusses parallel to and beside the drive, so a 14-foot wide head space over the drive is open for hoisting.

Consider installing overhead hopper-bottom bins in the receiving area for fast grain loadout into truck trailers and to assist blending grain for marketing. These bins also can hold grain for drying or feed processing. Loadout bins can be located just outside the receiving building near the dump pit or inside the building. On large farms, grain receiving and marketing loadout can conflict at harvest. Consider putting overhead loadout bins in a separate lane around the perimeter of the complex if hauling grain to a nearby processor or shipping conflicts during harvest. Size loadout bins to hold at least one semi-truck load (at least 1,000 bushels).

Grain centers with bucket elevators usually combine receiving and loadout because the bucket elevator discharge is directly above the grain receiving area. Inclined conveyors tend to separate receiving from loadout because the intake is farther from the discharge.

## Grain Receiving Capacity

With no dump pit, the grain-receiving conveyor must have high capacity to keep up with truck unloading during harvest. With a large dump pit, the conveyor's capacity can be as low as the harvesting capacity.

For transport vehicles of different capacities, size the capacity of the grain-receiving conveyor based on the time needed to unload the largest vehicle. In a two-vehicle system, the largest vehicle must travel to the grain center, unload, and return while the smallest vehicle is filled by the combine harvester.

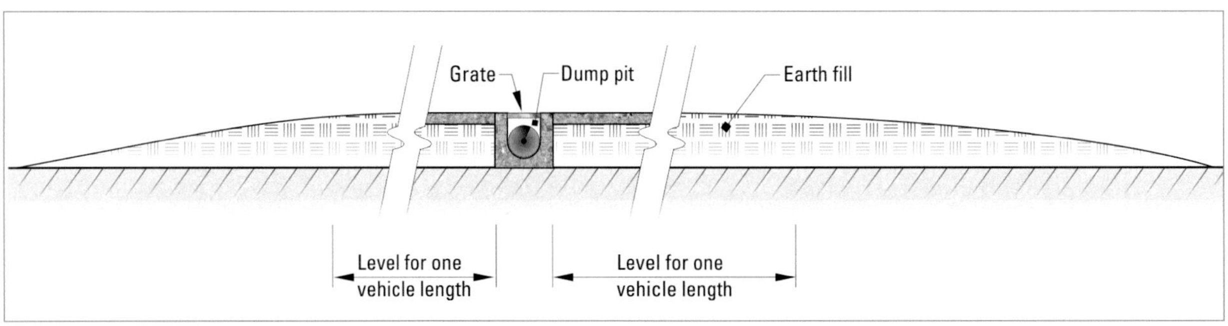


**FIGURE 3-5. Dump pit in earth fill above grade.**
Earth fill above grade for dump pit drainage. Build the ramp up to 4 feet above grade and at least 16 feet wide. Extend the ramp at least one vehicle length in each direction. Consider 8- to 9-foot width by 6- to 8-inch thick, reinforced concrete pavement at least one truck length each way from pit.

**Grain receiving capacity estimations for a two-vehicle system.**

See Figure 3-6 for definitions of the terms used in the example.

1. Estimate maximum harvest rate (MHR, bu/hr) and determine grain holding capacity of the largest (LV, bu) and smallest (SV, bu) transport vehicles. See the section on harvesting capacity on page 137.

2. Estimate travel time for the largest vehicle to go from the farthest field to the grain center (TTG, min), and to return to the same field (TTR, min). Empty vehicles may travel faster than loaded ones. With time in minutes, distance in miles, and speed in miles/hr, calculate estimated travel times using

$$\text{TTG or TTR} = 60 \times \text{distance} \div \text{speed}$$

3. Assume at least 5 minutes for miscellaneous activities (TMA, min) such as checking moisture content, starting conveyors, cleanup, and checking the dryer.

4. Calculate time to fill the smallest vehicle ($S_{fill}$, min)

$$S_{fill} = SV \times 60 \div MHR$$

5. The largest vehicle travels to the grain center, unloads, and returns while the smallest vehicle: filled by the combine harvester. Calculate the time available for the largest vehicle to unload at the receiving area (AUT, min):

$$AUT = S_{fill} - TTG - TMA - TTR$$

6. Unloading conveyor capacity (UC, bu/hr) can be as low as the harvest rate if a dump pit acts as a surge tank between loads. Calculate the minimum receiving pit capacity required for the unloading capacity to equal the harvest rate ($PC_{hr}$, bu):

$$PC_{hr} = LV - (MHR \times AUT \div 60)$$

If actual pit capacity ($PC_a$, bu) is greater than $PC_{hr}$, UC can be as low as the harvest rate. Larger unloading conveyor capacity allows more flexibility for changes in harvesting capacity or transport vehicles.

If $PC_a$ is less than $PC_{hr}$ or if there is no pit, calculate the unloading capacity (UC, bu/hr) ($PC_a$ = 0 if no pit):

$$UC = (LV - PC_a) \times 60 \div AUT$$

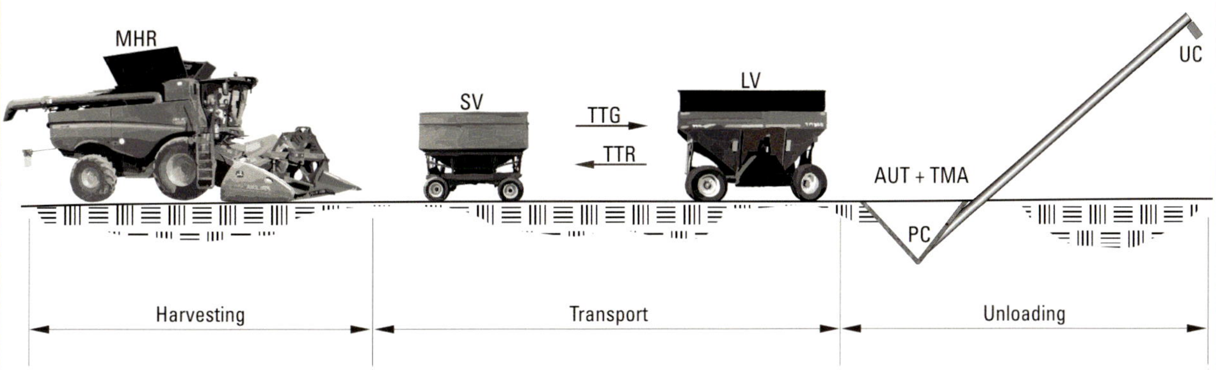


**MHR** = maximum harvest rate, bu/hr

**SV** = capacity of smaller transport vehicle, bu
**LV** = capacity of larger transport vehicle, bu
**TTG** = travel time for the largest vehicle to go to grain receiving, min
**TTR** = travel time for the largest vehicle to return to field, min

**AUT** = available unloading time, min
**TMA** = time for miscellaneous activities, min
**PC** = pit capacity, bu
**$PC_{hr}$** = pit capacity based on harvest rate
**$PC_a$** = actual pit capacity
**UC** = unloading conveyor capacity, bu/hr

**FIGURE 3-6. Definitions for the example of grain receiving capacity.**

**EXAMPLE 3-1.**

1. Determine required unloading capacity for a 400 bu/hr (MHR) harvest rate, one 500 bu (LV) transport vehicle, and one 300 bu (SV) transport vehicle.

2. The largest transport vehicle's speed averages 10 mph (both ways) from the farthest field 1 mile from the grain center.

$$TTG = TTR = 60 \text{ min/hr} \times 1 \text{ mile} \div 10 \text{ mph} = 6 \text{ min each way}$$

3. Assume 5 minutes for miscellaneous activities (TMA).

4. Calculate the time to fill the smallest vehicle:

$$S_{fill} = 300 \text{ bu} \times 60 \text{ min/hr} \div 400 \text{ bu/hr} = 45 \text{ min}$$

5. Calculate time available for the largest vehicle to unload:

$$AUT = 45 - 6 - 5 - 6 = 28 \text{ min}$$

6. Calculate minimum receiving pit capacity for unloading to keep up with harvesting capacity:

$$PC_{hr} = 500 \text{ bu} - (400 \text{ bu/hr} \times 28 \text{ min} \div 60 \text{ min/hr}) = 313 \text{ bu}$$

Calculate required unloading conveyor capacity:

If $PC_a$ is greater than 313 bu, UC can be as low as MHR, i.e. UC = 400 bu/hr.

If $PC_a$ is less than 313 bu, e.g., $PC_a$ = 250 bu:

$$UC = (500 \text{ bu} - 250 \text{ bu}) \times 60 \text{ min/hr} \div 28 \text{ min} = 536 \text{ bu/hr}$$

With no dump pit ($PC_a = 0$):

$$UC = (500 \text{ bu} - 0 \text{ bu}) \times 60 \text{ min/hr} / 28 \text{ min} = 1{,}071 \text{ bu/hr}$$

Example 3-1 shows that the capacity of the dump pit dramatically influences grain receiving capacity. Consider increasing the capacity of the unloading conveyor and/or the capacity of the receiving pit to meet future demand. The capacity of the unloading conveyor and the receiving pit may be difficult to change in the future.

One way to increase future capacity of the dump pit may be to build a 2- to 3-foot high approach and departure ramp to and from the dump pit area to increase the dump pit's wall depth. Also, adding fixed or flip-up 12- to 18-inch sidewalls or baffles on three sides of the pit for temporary additional holding volume can increase temporary and intermittent pit capacity.

## Roof Conveyors

Roof conveyors, shown in Figure 3-7, transfer grain to the top of one or more storage bins. With a bucket elevator, roof conveyors reduce the required leg height for gravity spouting, particularly for bins farther than one bin diameter away. Roof conveyors can be difficult to maintain unless catwalks are installed for better and safer access.

Roof conveyors are usually screw augers, but can be belt or chain-drag conveyors. Screw augers are usually enclosed by U-troughs, especially if large grain volumes are handled frequently. Augers mounted at the bin roof edge commonly fill large diameter bins, because portable conveyors cannot

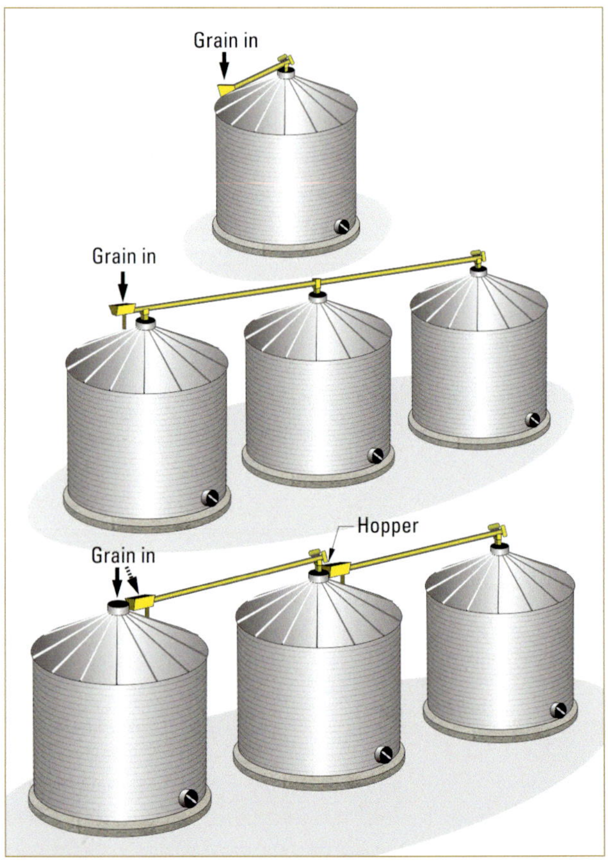


**FIGURE 3-7. Roof conveyor types.**

reach the roof peak. Size a roof conveyor for 5 to 10% more capacity than the one feeding it. Make sure the feeding conveyor shuts off if the roof conveyor stops. Check the roof structure to determine if the roof is strong enough to handle the load and vibration of a full, operating conveyor. Additional structural support frames from the sidewall or ground may be needed. Most bin manufacturers list the load capacity of their roofs, but if you are not sure if the roof will handle the load, check with the bin manufacturer.

While a roof conveyor to more than one bin fills an intermediate bin, some grain can carry over to the end of the conveyor and plug it. Install drop outs that permit carry-over grain to discharge into the last bin or through a spout on the end of the conveyor to a temporary storage or wagon.

## Grain Spreaders

Provide electrically-powered grain spreaders (levelers) in drying bins. When properly adjusted, they create a fairly level grain surface with fines spread throughout the grain mass. Without a spreader, fines tend to concentrate in the center and cause uneven drying, making aeration and grain cooling of the center grain mass difficult during the storage season.

*For dry grain storage bins*

- Electric spreaders may not be justified in storage bins less than about 24 feet in diameter. Consider a gravity-powered spreader, illustrated in Figure 3-8, or a grain cone to reduce packed grain and accumulated fines in the center of smaller bins. Multi-channel gravity grain spreaders are preferred.
- Another option is not to use grain spreading and instead unload/remove a portion of the grain (known as coring) during the bin filling process. Coring helps improve aeration airflow through the center grain mass by periodically removing the dense core of fines in the center of the bin. Coring the bin is done by operating unload conveyors periodically while the bin is being filled after several feet of new grain is added. Remove grain at regular intervals (daily or more often) during filling to remove the cone and invert the peak. Withdraw grain until the center 4- to 8-foot diameter (or 1/4 to 1/3 of the bin's diameter) surface is an inverted cone. Figure 3-9 shows this method, which works best for bins 18 feet in diameter and larger.
- Clean fines from the grain that is unloaded during coring, then recycle the cleaned grain into this bin or another bin when the bin is almost full. If cleaning cored grain is not desired, collect the cored grain until the bin is almost full, then recycle the cored grain back into the bin where it will be spread across the surface as a relatively thin top layer.

See the MWPS publication AED-20, *Managing Dry Grain in Storage*, for more information. Available at www.mwps.org.

**a. Gravity grain spreader for dry grain.**

**b. Multi-channel gravity grain spreader.**
*Photo courtesy of AgriDry LLC*

**FIGURE 3-8. Gravity-powered grain spreaders.**

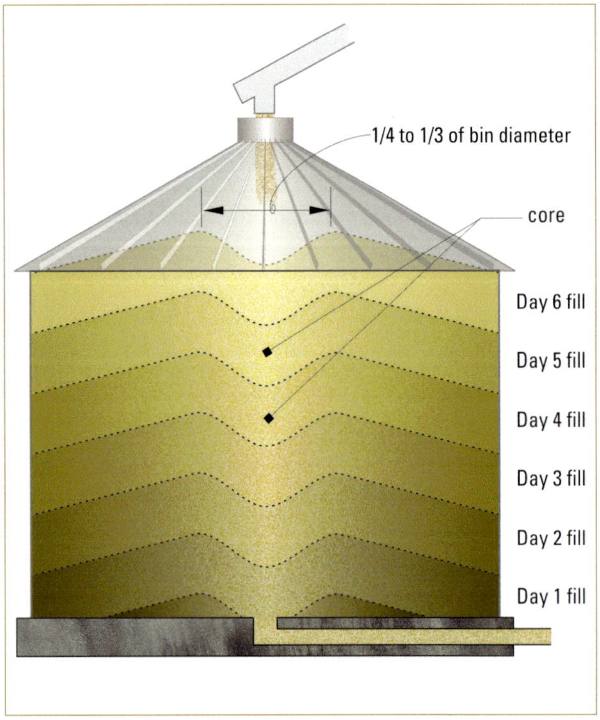


**FIGURE 3-9. Coring of the grain mass to remove fines in storage bins.**
Repeated withdrawals while the bin is filled removes most of the fines that accumulate in the center core. One withdrawal after the bin is filled suffices for smaller diameter bins.

## Grain Cleaning

Fines, foreign material, and broken kernels are grain handling problems. Kernels break during harvesting and handling. Cleaning grain before storing improves storability. But, unless fines cause serious storage problems, cash grain farmers with no market for the cleanings may lose money by cleaning grain. The broken grain, fines, and foreign material must be collected, perhaps stored, and fed to large size animals (mature cattle), sold, or discarded. (Fines may be used by livestock farmers or sold as feed, but care must be taken because of the potential for concentrating mycotoxin levels.)

Select a grain cleaner that collects and conveys screenings away with little or no attention. Some options for grain-cleaning equipment in a grain-handling system are shown in Figure 3-10, page 33.

*The most common locations for grain-cleaning equipment are*

- At receiving, before drying.
- After the dryer, just before delivery to storage.
- At loadout.

Cleaning is easier and more complete at low flowrates, such as, for example, at the output of a continuous-flow dryer. Collect screenings in a wagon or a special bin. Dry grain is more effectively cleaned than wet grain, and dry screenings are easier to store than wet ones. If you have an immediate use for wet screenings, it is better to clean ahead of drying to minimize use of extra fuel to dry the screenings. If grain is dried and stored in the same bin, clean the grain before drying.

Corn screenings can be good feed for livestock. Feed wet screenings within a day or two after collecting. You can give or sell screenings to local livestock farmers. Take care when feeding fines from damaged grain. Moldy, soft, partially rotted grain particles are screened with the fines and may include toxins (aflatoxin in corn) harmful to livestock. Toxins can be greatly intensified in screened materials.

## Grain Loadout

Storage bins can unload directly to transport vehicles by a portable auger or by side discharge.

Figure 3-11, below, shows a portable, inclined screw auger fed by an underfloor auger. In this method, a sweep auger removes the grain that does not flow out by gravity. It feeds the underfloor auger which transfers grain to an external inclined conveyor or other conveyor.

Side discharge is a convenient and efficient way to unload grain bins; see Figure 3-12 for an example of a bin with an external side discharge. Special structural design and support needs to be considered to avoid any potential structural damage or wall collapse that can be caused by grain flowing out of one side of a bin and shifting the load to the opposite side of the bin. Larger sized bins can have multiple side discharges. A slide gate is used to control the grain flow.

**FIGURE 3-11. Inclined conveyor bin loadout from sweep auger and underfloor auger.**

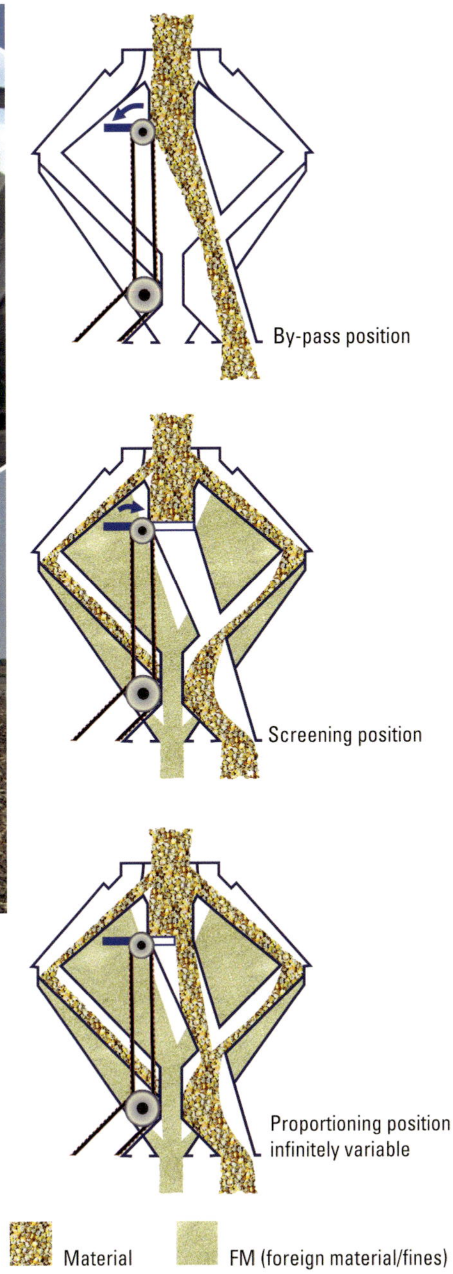


**a. Bucket elevator and gravity screen cleaner, with detail.**
*Photo courtesy of Intersystems (GSI). Detail adapted from GSI diagram.*

**b. Suction air grain cleaner for separation of bee's wings.**
*Photo courtesy of KLEENair*

**c. Rotary screen cleaner.**
*Photo courtesy of Farm King*

**FIGURE 3-10. Grain cleaners.**

**FIGURE 3-12. Grain bin with external side unload (a) and inside baffle (b) allowing grain to flow out while providing wall support.**
*Photos courtesy of Agri-Systems, Inc.*

## Grain Bin Cleanout—Sweep Unloaders

Grain loadout is fastest when grain flows by gravity to the in-floor sump intake in the center of a bin. In larger diameter bins additional sumps can be opened once grain stops flowing by gravity out of the center. An inverted cone or series of cones of grain remain that must be removed by an above-floor sweep unloader. If bins are loaded and unloaded multiple times during the harvest and storage season, leave the grain that does not flow to the center sump in the bin to form a hopper. At the end of the season, remove the grain hopper with a sweep unloader to avoid leaving old grain into the following season, which can cause spoilage problems.

However, be aware that when the grain hopper is not removed periodically, grain fines will filter into the grain hopper surface, making the unremoved grain hopper denser, as fines will fill the inter-kernel void space. After several bin loads have been unloaded by gravity only, the density of the unremoved grain hopper makes cleanout by a sweep unloader difficult. If a grain bin is to be completely unloaded frequently, or if each bin load is unloaded by a sweep unloader maintaining unload capacity at close to the system's gravity flow capacity, then the sweep unloader must be more robust in its design

For grain storage bins that are unloaded only once per season, the bin sweep auger's primary function is removing the grain hopper that remains after gravity flow stops instead of matching unload conveying capacity. In this situation, the diameter of the sweep auger screw should be relatively well matched to the capacity of the underfloor unload auger, but not oversized to match that of the downstream unload conveyors.

> **For example:** If the bin has an 8-inch discharge auger, the sweep unloader should be at least 7-inches in diameter. Cleanout capability of the sweep auger is more important than its speed. After two to four passes around the bin, the sweep auger should remove all but the final 2 to 4 inches of grain.

To provide good cleanout of bins, the sweep unloader must have a back shield and must have excellent drive wheel traction with the bin floor. This minimizes the number of passes required to clean the bin floor.

Sweep unloaders must be capable of bin cleanout without workers entering the bin to assist the machine during cleanout. Manufacturers of sweep-unloader equipment provide warning statements, labels and decals on their equipment explicitly stating that personnel are not to enter bins when sweep unloaders are powered. Many workers suffer injuries by entering bins and trying to assist sweep unloaders during bin cleanout.

Figure 3-13 illustrates one type of sweep unloader currently available for bin cleanout.

## Grain Bin Cleanout—Chain-Paddle Unloaders

Recently, a cleanout device for round steel bins that does not use rotating auger flighting has been developed and introduced for cleaning out bins in farm and commercial operations. This new device greatly improves the safety of the cleanout process for flat-bottom bins. The round bin cleanout device shown in Figures 3-14 and 3-15, uses rubber paddles on a chain drive under a steel shroud housing to sweep the grain to the sump at the bin's center. It moves due to heavy center wheels powered by electric motors.

The back side of the housing has a rubber edge that drags the floor, so the rubber paddles that brush the floor also slide against the back inside wall of the conveyor housing, cleaning all grain ahead of the back wall housing with a rubber bottom edge, much like the rubber wiper on the bottom edge of a household storm door.

The unique design allows the unit to start underneath the inverted grain hopper and clean the bin floor in one pass with virtually no grain left in its path, except against the bin wall. This device is much safer if personnel were to enter the bin while the device is operating, as it tends to push large objects that are not grain to the side, away from the paddles.

Horizontal drag conveyors, like those shown in Figures 3-16 and 3-17 can return grain to receiving/loadout, for transfer to trucks or to other locations in the grain center at high capacities.

If dry grain is frequently loaded out while wet grain is received, consider separating the two areas. Usually receiving and loadout in the same drive do not conflict until grain volumes exceed about 100,000 bushels per year.

## Grain Impact and Velocity

Dropping grain from excessive heights into bins or down steep gravity spouts can cause serious grain breakage. Steep gravity spouts often occur above loadout or low bins near the bucket elevator. Cushion boxes and in-line flow retarders reduce grain velocity and damage, as well as spout wear but are not completely effective. Avoiding steep spout slopes and using an appropriate slope angle will keep grain damage low. Cushion boxes like those shown in Figure 3-18 should have removable end caps for cleanout maintenance to avoid an accumulation of moldy grain and rust/corrosion of the spouts. The self-cleanout cushion box shown in Figure 3-18b is the preferred design.

**FIGURE 3-13. Typical sweep unloader with back shield and a drive wheel.**

*Photo courtesy of Sukup Manufacturing Co.*

**FIGURE 3-14. New chain drag cleanout conveyor with rubber paddle and powered drive wheels.**
Cleans grain bin floors in one pass.

*Photo of DAAY Bin Power Sweep courtesy of Sioux Steel Co.*

**FIGURE 3-15. Rubber paddles shown dragging across floor; chain and paddles return on a horizontal divider; floor slide brackets control the height of paddles to minimize floor pressure.**

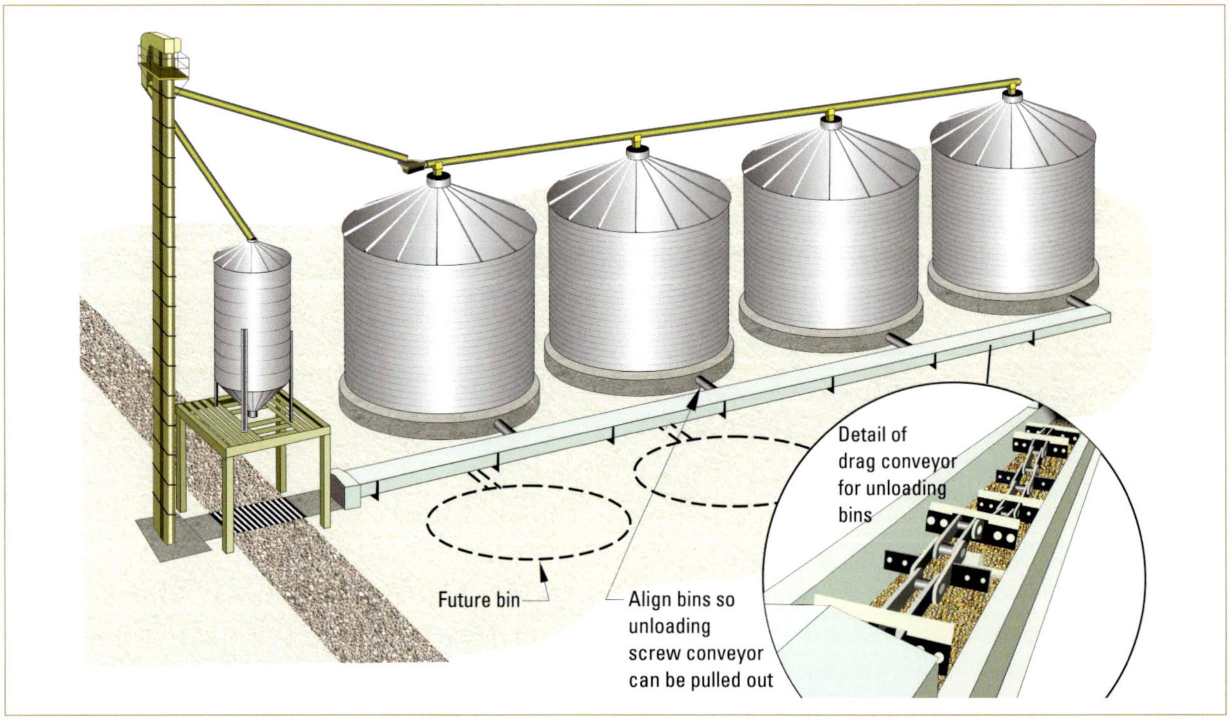


**FIGURE 3-16. Loadout to horizontal drag conveyor.** Offset the unloading screw conveyors if bins are added on the other side of the drag conveyor so screw conveyors can be easily removed for service.

**FIGURE 3-17. Example of loadout to a chain-conveyor loop system.**

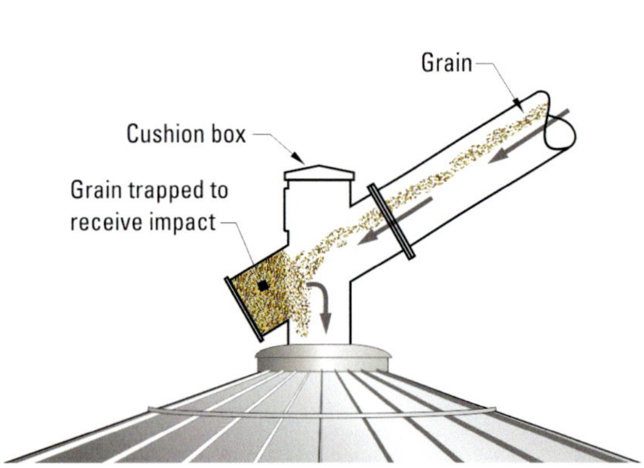


**a. Capped spout piece holds grain quantity for falling grain to strike.** Grain remains in the box after grain transfer stops. If this grain gets wet from rain or condensation, it may become moldy and attract insects.

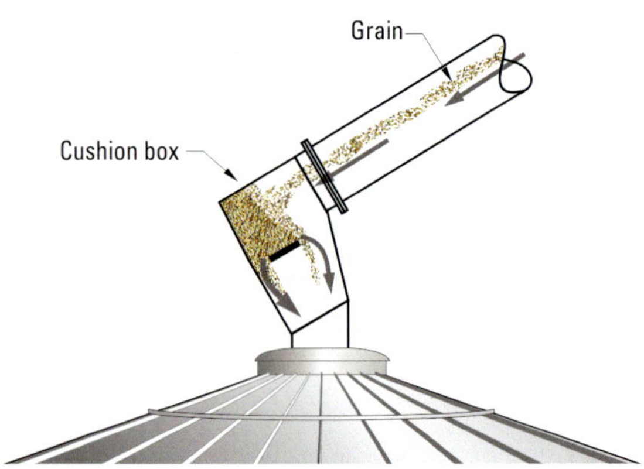


**b. Self-cleaning type cushion box allows for grain to empty.**

**FIGURE 3-18. Examples of cushion boxes.**

# GRAIN CONVEYING EQUIPMENT

Table 3-2 summarizes common conveyor types, their power requirement, advantages and disadvantages, specific applications follow. Many grain handling systems have more than one conveyor type and conveying capacity.

**TABLE 3-2. Grain conveyors.**

| Type of conveyor | Horsepower requirement per unit capacity | Advantages | Disadvantages |
|---|---|---|---|
| Auger or screw conveyor | Low to medium with dry grain and medium to high with wet grain | • Simple, widely available in many sizes<br>• Initial cost is low<br>• Available for horizontal, inclined, vertical applications; portable, wheeled, or fixed<br>• Adapted to most grain and feed materials<br>• Useful as mixer, flow meter, force feeder, or agitator<br>• Can have multiple inlets and outlets | • High torque and power required in wet grain<br>• Medium to heavy wear<br>• Noisy if not supported and/or not operated at rated capacity<br>• High grain damage if not operated at rated capacity<br>• Can be used only in straight lines<br>• Can be dangerous |
| Belt | Low | • Good for long distances<br>• Low power requirement<br>• Quiet<br>• Gentle material handling with least handling damage. Useful for edible food-grade grains, soybeans and seeds<br>• Capacity only affected by grain weight<br>• Self cleaning<br>• High efficiency | • Limited in angle of elevation<br>• Expensive<br>• Belt maintenance<br>• Difficult to enclose<br>• Not made for easy intermediate discharge points |
| Bucket | Low | • Efficient, compact<br>• Low maintenance<br>• Quiet<br>• High capacity for vertical lift<br>• Reliable and adaptable to automation<br>• Easily cleaned | • Difficult to erect and change (increase capacity)<br>• Expensive<br>• Grain damage high for large drop heights<br>• Elevator head servicing is difficult<br>• Can emit dust unless attached to dust aspirators |
| Bulk or mass flow | Low | • Good for long distance<br>• Low power requirement<br>• Quiet<br>• Low maintenance<br>• Little grain damage | • Expensive<br>• Material carryover |
| Flight | Low to medium | • Low cost<br>• Flexible<br>• Useful for materials other than grain | • Noisy<br>• Heavy wear<br>• Dense materials difficult to convey |
| Pneumatic | High | • Flexible installation<br>• Easily cleaned<br>• Convenient grain delivery to many locations<br>• Ideal for hot grain transfer from dryer to dryeration bin | • High power requirement<br>• Creates dust, usually requires separation equipment |

## Auger Conveyors

Screw augers convey grain at any angle from horizontal to vertical, can be fixed or portable, are relatively low cost, and come in many diameters and capacities. Large-diameter augers have high capacity but require higher horsepower. Compared with other conveyors, augers generally have low capacity per horsepower, especially for wet grains. For example, compared with dry shelled corn, high-moisture shelled corn requires nearly twice the horsepower for half the capacity. High-capacity farm augers are usually powered by tractor PTO.

The simplest tube augers have bearings only at the drive end. When allowed to run empty, flighting wear, and noise are high. Better tube augers have bearings at both ends and at intermediate points depending on length. Usually they have heavier gauge flighting and casings. Properly spaced bearings are quieter and can prevent the flighting from wearing on the enclosure when empty.

Contrary to common belief, augers are not a primary source of grain damage if they are operated full at their rated speed. Damage can occur when grain is pinched between the auger flighting and casing. Prevent pinching by selecting augers with the clearance between the auger flighting and casing either greater or less than the average grain diameter being conveyed. Figure 3-19, illustrates the components of a typical tube auger and presents the terminology used to describe those components.

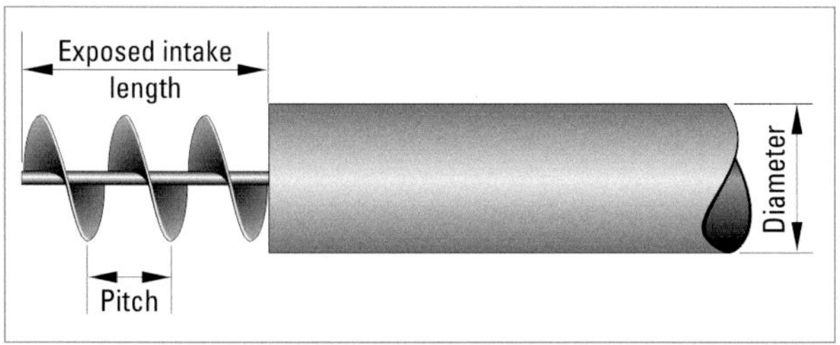


**FIGURE 3-19. Tube auger definitions.**

***Maximum*** *recommended auger speeds are*

- 800 revolutions per minute (rpm) for 4-inch augers.
- 600 rpm for 6-inch augers.
- 450 rpm for 8-inch augers.

Consider a larger auger and run it more slowly to decrease wear, increase power efficiency, and reduce grain damage. For a continuous-flow dryer with an output of 600 to 800 bushel per hour, choose a 6- or 8-inch auger rather than a 4-inch one, and run it more slowly. Slow speeds may require a gear reducer, which is a good investment over the life of the equipment. Because they can damage the seed coat, augers are not generally used for handling edible beans or seed grain.

U-trough augers are preferred for horizontal conveying of large volumes. Consider them for hard-to-maintain locations, such as overhead augers and augers in dump pits. They operate at slower speeds, which increases auger life and may reduce grain damage, but they are more expensive than other auger types.

U-trough augers are generally used for horizontal to shallow-inclined installations when higher capacity or longer use is required on large farms and in commercial elevator installations. U-trough augers are a lower cost conveyor compared to drag or belt conveyors and are usually specified when conveyor sizes of 10-inch diameter or larger are needed. U-trough conveyors typically operate from 150 to 300 rpm at fill capacities of at least 45 to 50%.

U-trough augers do not work well at inclines over 15 degrees because the enclosure does not hold grain in the flighting. The top cover is removable for repair. U-trough augers may require more support than tube augers. Catwalk support systems, with proper safety measures, are common for overhead installations.

Auger capacity and horsepower are affected by diameter, pitch, speed (rpm), exposed intake length, incline, and grain properties. Auger length affects power requirement, but not capacity.

Auger capacity increases with diameter. A pitch length equal to flight diameter (single pitch) gives the best performance under most conditions. Most farm augers are single pitch.

Capacity and required horsepower increase with speed. The increase is less for an inclined auger or short intake length. Capacity increases until the grain has difficulty entering the auger because the intake flighting throws it back out, usually at speeds beyond normal operation. Grain damage and auger wear tend to increase with speed, especially if the auger is not run at full capacity.

Capacity and power requirements increase with exposed intake length. At some point, more exposed intake increases horsepower but not capacity. Optimum exposed intake length at full capacity and rated rpm is generally 2 to 3 times the auger's diameter.

For a given horsepower, increased auger incline decreases capacity. Horsepower increases from horizontal to a maximum at inclines between 45 and 60 degrees. Above a 60-degree incline, capacity loss reduces horsepower. At 90 degrees, the power required is about the same as for horizontal, but capacity is much less.

Higher grain moisture content reduces capacity and increases power requirements because wet grain is heavier and has more friction. Expect auger capacities to be reduced by 40% with moist grain.

Grain type influences capacity and power requirements. An auger's capacity for wheat, grain sorghum, oats, barley, and rye is a little less than for corn. With soybeans, an auger has lower capacity and higher power requirements than with corn.

Substantial grain depth over the auger intake, as in bulk tanks, increases auger horsepower needs, especially with long exposed intakes. Half pitch flights (pitch = 1/2 auger diameter) run at rated rpm can reduce the amount of grain entering the auger, so horsepower is reduced.

Table 3-3 gives capacities and power requirements of single-pitch augers handling dry shelled corn. Adjust Table 3-3 values for different auger speeds with Table 3-4 and for wet corn with Table 3-5.

The values from Table 3-3 do not include drive-train losses; increase by 10% to determine required motor power. If an auger occasionally handles wet grain, size the motor for wet-grain power requirements. Sizing motors, pulleys, and V-belts is discussed more fully in the Appendix of this handbook.

**TABLE 3-3. Estimated auger capacity and power.**

- For dry (14% maximum moisture content) corn.
- Values for 4-inch and 6-inch augers are based on an extensive set of data.
- Values for 8-inch and 10-inch augers are based on a limited set of data.
- Values for 12- to 16-inch augers were extrapolated.
- Actual auger performance may vary, so use manufacturer's data for designing screw conveying systems.

Multiply dry corn values by 0.6 for wet corn capacity. Use the table values for wheat, grain sorghum, oats, barley, and rye because actual values are only slightly less. For soybeans, multiply capacity by 0.75 and power by 1.10. Multiply bu/hr by 50 for approximate lb/hr capacity for meal or concentrate feed.

| Auger dia., in. | Auger speed[1] rpm | Incline angle 0° | | 25° | | 35° | | 45° | | 90° | |
|---|---|---|---|---|---|---|---|---|---|---|---|
| | | bu/hr | hp/10' | bu/hr | hp/10' | bu/hr | hp/10' | bu/hr | hp/10' | bu/hr | hp/10' |
| 4 | 900[2] | 560 | 0.6 | 500 | 0.9 | 480 | 0.9 | 450 | 1.0 | 270 | 0.8 |
| 6 | 600 | 1,500 | 1.0 | 1,350 | 1.5 | 1,290 | 1.6 | 1,190 | 1.6 | 710 | 1.3 |
| 8 | 450 | 2,210 | 1.4 | 1,990 | 2.2 | 1,890 | 2.2 | 1,760 | 2.3 | 1,050 | 1.8 |
| 10 | 360 | 3,300 | 2.0 | 2,970 | 3.1 | 2,830 | 3.2 | 2,620 | 3.2 | 1,570 | 2.5 |
| 12 | 300 | 4,520 | 2.5 | 4,070 | 3.9 | 3,870 | 4.0 | 3,590 | 4.0 | 2,150 | 3.2 |
| 14 | 260 | 6,230 | 3.4 | 5,610 | 5.3 | 5,340 | 5.4 | 4,950 | 5.5 | 2,960 | 4.3 |
| 16 | 225 | 8,040 | 4.4 | 7,240 | 6.8 | 6,870 | 7.0 | 6,390 | 7.1 | 3,820 | 5.6 |

[1]Auger speeds for 3,600 in/min flighting velocity along auger length (theoretical grain velocity) for all diameters.
[2]4-inch auger at 900 rpm vibrates excessively. The 900 rpm values are for converting with Tables 3-4 and 3-5.

**TABLE 3-4. Conversions for auger speed.**

For converting from Table 3-3 for different auger speeds. You can interpolate between speeds with reasonable accuracy. These values are for dry corn; convert first for speed, then for moisture content using Table 3-5.

| Speed relative to Table 3-3 % | Multiplier Capacity | Power |
|---|---|---|
| 125 | 1.17 | 1.23 |
| 100 | 1.00 | 1.00 |
| 75 | 0.79 | 0.76 |
| 50 | 0.56 | 0.51 |
| 25 | 0.29 | 0.26 |

**TABLE 3-5. Conversions for wet (25%) corn.**

Wet corn capacity = 0.6 × dry corn capacity (Table 3-3). This table converts power requirements from Table 3-3 for wet corn. If the auger also operates at a different speed, first convert for speed, Table 3-4, then for moisture.

| Speed relative to Table 3-3 % | Incline angle 0° | 25° | 35° | 45° | 90° |
|---|---|---|---|---|---|
| | - - -hp multiplier - - | | | | |
| 125 | 2.6 | 1.9 | 1.8 | 1.7 | 1.0 |
| 100 | 2.8 | 2.0 | 1.9 | 1.8 | 1.0 |
| 75 | 3.1 | 2.2 | 2.1 | 2.0 | 1.0 |
| 50 | 3.8 | 2.7 | 2.6 | 2.5 | 1.0 |
| 25 | 3. 5 | 4 8 | 3.5 | 3.3 | 1.0 |

#### EXAMPLE 3-2.

What is the estimated capacity and required motor size for a 51-foot long, 6-inch diameter auger operating at 450 rpm with a 35-degree incline?

#### SOLUTION 3-2.

From Table 3-3, a 6-inch auger running at 600 rpm with a 35-degree incline, conveys 1,290 bu/hr of dry corn and requires 1.6 hp per 10 feet of auger length.

Determine relative operating speed:

$$(450 \text{ rpm} \div 600 \text{ rpm}) \times 100 = 75\%$$

From Table 3-4, the relative capacity multiplier is 0.79, and the relative power multiplier is 0.76 when the relative speed is 75%. Calculate the relative capacity and power requirement.

Capacity at 450 rpm =
$0.79 \times 1{,}290$ bu/hr = 1,019 bu/hr

Horsepower per 10' at 450 rpm =
$0.76 \times 1.6$ hp/10' = 1.22 hp/10'

Total horsepower =
$(51' \div 10') \times 1.22$ hp/10' = 6.22 hp.

From Table A-17, select a 7-1/2 hp electric motor or a 12 hp gasoline engine.

#### EXAMPLE 3-3.

Determine the estimated capacity and horsepower for conveying wet corn with the auger of Example 3-2.

#### SOLUTION 3-3.

From Example 3-2, the adjusted auger capacity is 1,019 bu/hr and total conveying hp is 6.22. Adjust the capacity and required power for moisture with Table 3-5. Multiply capacity by 0.6. From Table 3-5, hp multiplier is 2.1 when the relative speed is 75% at 35-degree incline.

Capacity for wet corn at 450 rpm =
1,019 bu/hr $\times$ 0.6 = 611 bu/hr.

Total hp for wet corn at 450 rpm =
6.22 hp $\times$ 2.1 = 13.06 hp.

From Table A-17, select a 15 hp electric motor or 23hp gasoline engine.

#### EXAMPLE 3-4.

Determine the capacity and horsepower if conveying soybeans with the auger in Example 3-2.

#### SOLUTION 3-4.

From Example 3-2, the adjusted auger capacity is 1,019 bu/hr and total conveying hp is 6.22. Adjust the capacity and power values for soybeans by multiplying capacity by 0.75 and hp by 1.10, from Table 3-3.

Capacity for soybeans at 450 rpm =
$0.75 \times 1{,}019$ bu/hr = 764 bu/hr.

Total hp for soybeans at 450 rpm =
$1.10 \times 6.22$ hp = 6.84 hp.

From Table A-17, select a 7-1/2 hp electric motor or a 12 hp gasoline engine.

Ease of maintenance is important. Bin unloading augers must be removed for repair or maintenance. Locate them with plenty of room to pull the auger completely out of the bin without having to cut the auger. Place catwalks beside overhead conveyors for easier and safer access. Covered U-trough augers are easier to maintain than tube augers.

Augers are a major farm hazard. Manufacturers put safety shields on exposed auger drive assemblies and grates over auger intakes to keep hands, feet, and clothing from contacting the auger flighting or being caught between the flighting and housing. Do not remove safety shields. **When children are around, never start any machinery without being sure they are not in danger.**

## Belt Conveyors

Portable belt conveyors, such as the one shown in Figure 3-20, handle grain gently, but are expensive and generally unsuitable for steep inclines. They have higher capacity per horsepower than other conveyor types (that is, they are more power efficient). Portable belt conveyors with improved design are being used increasingly to move delicate specialty crops, edible beans, and seed. See Table 3-6 for typical capacities of horizontal belt conveyors which are more typically used on commercial facilities than on-farm.

## Bulk- or Mass-Flow Conveyors

Bulk- or mass-flow conveyors are common in commercial elevators and for seed handling and are becoming more common on large farm systems. They convey across the top or under the bottom of storages and occasionally elevator pits. They are usually more expensive than other types of conveyors, but they are very power efficient (high capacity per horsepower) and reliable. The housings for bulk-flow conveyors are usually rectangular, but U-trough or round tube designs also are available. The flights are usually 8 to 12 inches wide and often only 3/4 to 1-1/2 inch high. These conveyors are called bulk flow because the shallow flight scraping along the housing bottom carries grain three to five times as deep as the flighting. The grain between the flights carries along the grain above. The flights are usually high density plastic or composition material precisely fit to the conveyor bottom and sides for excellent conveyor cleanout.

**FIGURE 3-20. Portable tubular belt conveyor.**
*Photo courtesy of CrustBuster Speed King Inc.*

**TABLE 3-6. Typical capacities for horizontal belt conveyors.**

Belt speed = 100 fpm (feet per minute); uniformly loaded.
Capacity depends on belt speed.
Capacity at your belt speed = table value × your belt speed ÷ 100.

| Belt width, in | Troughed belt | Flat belt | Maximum belt speed |
|---|---|---|---|
| | bu/hr at 100 fpm | | fpm |
| 12 | 370 | 183 | 350 |
| 14 | 508 | 228 | 400 |
| 16 | 675 | 305 | 450 |
| 18 | 833 | 373 | 450 |
| 20 | 1,074 | 482 | 450 |
| 24 | 1,575 | 707 | 600 |

*Source: University of California, Agricultural Circular 517.*

Rectangular conveyors are used mainly for inclines up to 15 degrees, although some have shallow, horseshoe-shaped flights for vertical conveying. Special tubular designs have paddles filling most of the tube for conveying at inclines up to vertical.

Avoid bulk flow conveyors for hot grain from a dryer in cold weather because moisture condenses in the conveyor. Conveyors exposed to weather must be rain tight to keep water out of equipment and storages.

See Table 3-7 for capacity and power requirements for horizontal bulk flow conveyors at 200 fpm chain speed.

*To adjust capacity and horsepower requirements for other depths and speeds use:*

- Capacity adjustment for conveyor depth:
  Bu/hr =
  (bu/hr for conveyor width from Table 3-7) ×
  (available conveyor depth/conveyor width)
- Capacity adjustment for chain speed:
  Bu/hr =
  (capacity, bu/hr from Table3-7) ×
  (chain speed, fpm ÷ 200 fpm)
- Horsepower adjustment for depth and chain speed:
  Hp/10' =
  (hp/10' from Table 3-7) × (adjusted bu/hr) ÷
  (bu/hr from Table 3-7)

**TABLE 3-7. Typical bulk-flow conveyor capacity and power.**

**For horizontal conveyors only.** Chain speed = 200 fpm. Conveying depth equals width, and available conveying area is 85% full.

| Conveyor width and depth, in | Capacity bu/hr | Power required hp/10' |
|---|---|---|
| 8 | 3,840 | 1.0 |
| 9 | 4,860 | 1.2 |
| 10 | 6,000 | 1.4 |
| 11 | 7,260 | 1.6 |
| 12 | 8,640 | 1.8 |
| 13 | 10,140 | 2.1 |

*From Buhler Conveyors, Hutchinson Corp, and Berreau literature. Corp.*

**EXAMPLE 3-5.**

Determine the capacity of a conveyor that is 9 inches wide and 6 inches deep.

**SOLUTION 3-5.**

From Table 3-7, conveyor capacity is 4,860 bu/hr for width = depth = 9" and chain speed = 200 fpm. To convert conveyor depth to 6":

Bu/hr = 4,860 bu/hr × 6" ÷ 9" = 3,240 bu/hr

**EXAMPLE 3-6.**

Adjust the conveyor from Example 3-5 for 150 fpm chain speed.

**SOLUTION 3-6.**

From Example 3-5, capacity is 3,240 bu/hr at 200 fpm. Convert to 150 fpm chain speed:

Bu/hr =
3,240 bu/hr × 150 fpm ÷ 200 fpm = 2,430 bu/hr.

**EXAMPLE 3-7.**

What horsepower is required for the conveyor in Example 3-6?

**SOLUTION 3-7.**

From Example 3-6, adjusted capacity is 2,430 bu/hr. From Table 3-7, capacity is 4,860 bu/hr, and power required is 1.2 hp/10' for the original conveyor. Adjusted power requirements are:

hp/10' =
1.2 hp/10' × 2,430 bu/hr ÷ 4,860 bu/hr = 0.6 hp/10'

## Flight Conveyors

Double-chain flight conveyors, like those for ear corn or baled hay, are low cost. Capacity decreases as flight depth or speed decreases and as flight spacing or conveyor incline increases. Although not designed for shelled corn, flight conveyors can convey wet corn to a holding or drying bin.

The flights are usually 17 to 21 inches wide, 2-1/2 to 3 inches high, and have speeds of 150 to 200 fpm. The conveyors are typically 25 to 60 feet long. Shelled corn and small grain capacity is about 1,000 to 2,000 bu/hr at a 15- to 35-degree incline. The motors used in flight conveyors range between 3 to 7-1/2 horsepower.

Research is not available on grain damage in flight conveyors, but it is believed to be low with flights fitting well inside the conveyor housing. However, a worn conveyor may seriously damage dry grain, and some grain damage occurs in flight conveyors because of rubbing action and possible pinch points between conveyor components. The damage is usually less than that which occurs with screw or pneumatic systems.

## Bucket Elevators

Vertical bucket elevators (belt and cup elevator legs) are often the hub of convenient, high-capacity grain centers. They elevate almost all the grain; while other conveyors are used to convey grain to and from the bucket elevator.

Bucket elevators handle wet and dry grain, whole and ground grain, meals, and animal feeds. Capacities and power requirements change little with grain moisture content. Power requirements are low compared with those of other conveyors, particularly for wet grain. They are quiet, reliable, have little wear, and have a long service life.

A single-bucket elevator, as shown in Figure 3-21, can have more than one function. Occasionally, two functions conflict.

> **For example:** Transferring wet grain from receiving to wet holding versus transferring dry grain from a continuous-flow dryer to storage at the same time. A surge bin can collect dry grain while the bucket elevator is conveying wet grain, or a second bucket elevator or other conveyor can be installed so the two functions can be handled at the same time. With one bucket elevator for multiple functions, consider a warning light or electrical lockout to remind or prevent the operator from routing grain into the wrong bin, for example, sending wet grain into dry storage.

The drying method may determine if one or two bucket elevators are needed. One bucket elevator is usually enough for batch drying processes: receiving wet grain, transferring dry or hot grain from the dryer if required, handling ground feed, and loadout of dry grain for marketing and processing. With a high-speed automatic batch dryer, consider placing the wet-holding bin over the dryer for fast gravity refill, or lift wet grain to the dryer with an inclined conveyor so the controls for the bucket elevator can be set for automatic transfer of dry or hot grain when the batch is finished. Manually control the bucket elevator for wet receiving.

A continuous-flow dryer usually requires two dedicated conveyors, typically two bucket elevators, illustrated in Figure 3-22. Use a high-capacity bucket elevator (usually 3,000 to 6,000 bu/hr) for wet corn receiving for dryer refill. If there is no overhead gravity-fill wet holding bin, and for dry grain loadout for marketing or processing. The second, lower-capacity elevator is usually matched to the maximum dryer output (typically 500 to 1,500 bu/hr). It conveys dry grain to final storage or hot grain to bin cooling or dryeration bins. A popular alternative is to use a pneumatic conveying system for hot grain transfer.

With two batch dryeration bins, the higher capacity bucket elevator may be taller and spouted to all bins for wet receiving, dryeration transfer, and dry loadout. The lower capacity leg is spouted only to the batch dryeration bins. Consider a gravity surge bin or separate conveyor to transfer wet grain to the dryer to avoid conflict.

With one continuous-flow dryeration bin, a second low-capacity leg is needed with a distributor that transfers dried and cooled grain to storage bins. If properly located close to the main leg, the dry-grain leg from the dryeration bin

**FIGURE 3-21. Grain center with single-bucket elevator.** A single-bucket elevator setup is common with a batch dryer. The same bucket elevator receives wet grain, discharges to the wet-holding bin, moves dried grain to dryeration or storage from the dryer in sequential order.

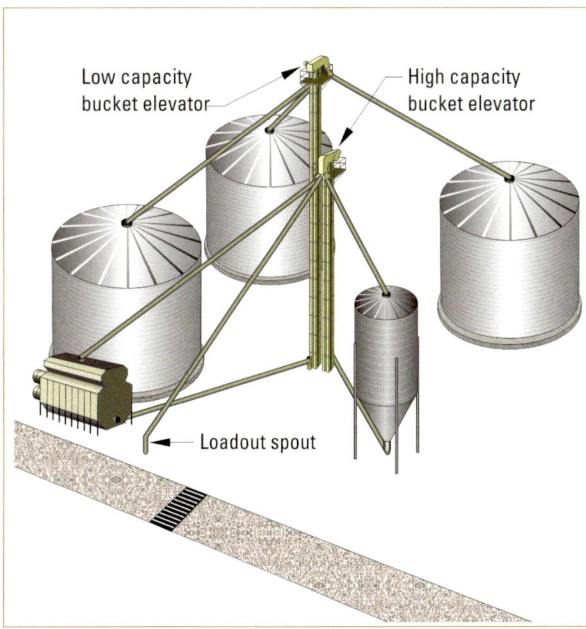


**FIGURE 3-22. Grain center with two bucket elevators.** A two-bucket elevator setup is common with a continuous-flow dryer. A high-capacity bucket elevator receives wet grain, discharges to dryer or wet-holding bin, moves grain from wet holding to dryer, and discharges for loadout. A low-capacity bucket elevator moves dried grain to dryeration or storage.

can transfer dry grain into the main distributor and route the grain to all storage bins, allowing both main and dryeration legs to use a common distributor.

Bucket elevators have a small footprint, but the support system can require more space, block passage, and restrict growth. Guying cables require more area than do tower support systems. Figure 3-23 details typical support systems for bucket elevators. Guying cables are located at 90-degree angles to each other. Consult the supplier of the bucket elevator for the required number and location of guying cables. Towers around the bucket elevator can be used to support it instead of guy cables. Towers usually take up 64- or 100-square feet of space and reduce obstructions, but they cost more than guying cables. Towers or cages around a bucket elevator also permit including safety features such as rest platforms and stairs, as seen in Figure 3-23b.

Figure 3-24 provides a schematic of a typical bucket elevator. Nearly all farm bucket elevators are centrifugal discharge, double-leg, with buckets mounted on a belt. Material discharges by centrifugal force as the buckets pass over the head wheel. To reduce impact damage, use slow-speed, gravity discharge for seeds and edible beans.

Intake hoppers can be on the up leg or down leg of the bucket elevator; however, upside hoppers fill cups directly, whereas down-leg hoppers cause cups to dig or scoop the grain out of the boot. Upside intake hoppers are typically used for grains. Downside intake work best for meals. Input to a hopper can be a gravity drop from a pit, spout, or conveyor.

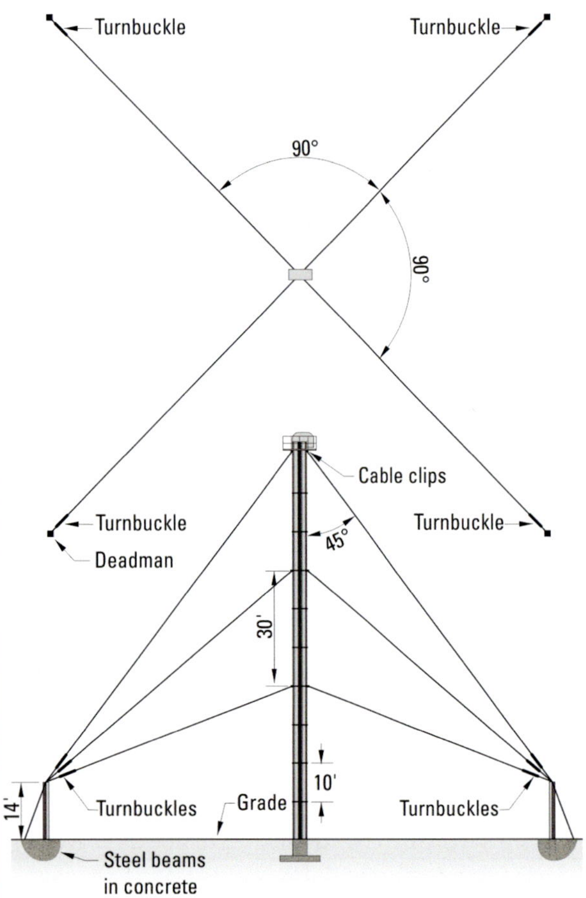


**a. Guying cable support for a bucket elevator.**
Have cables placed by the bucket elevator supplier.
*Adapted from Sweet Manufacturing.*

**b. Bucket elevator supported by a tower with stairs and resting platforms.**
*Photo courtesy of Farm Science Review, The Ohio State University.*

**FIGURE 3-23. Support systems for bucket elevators.**

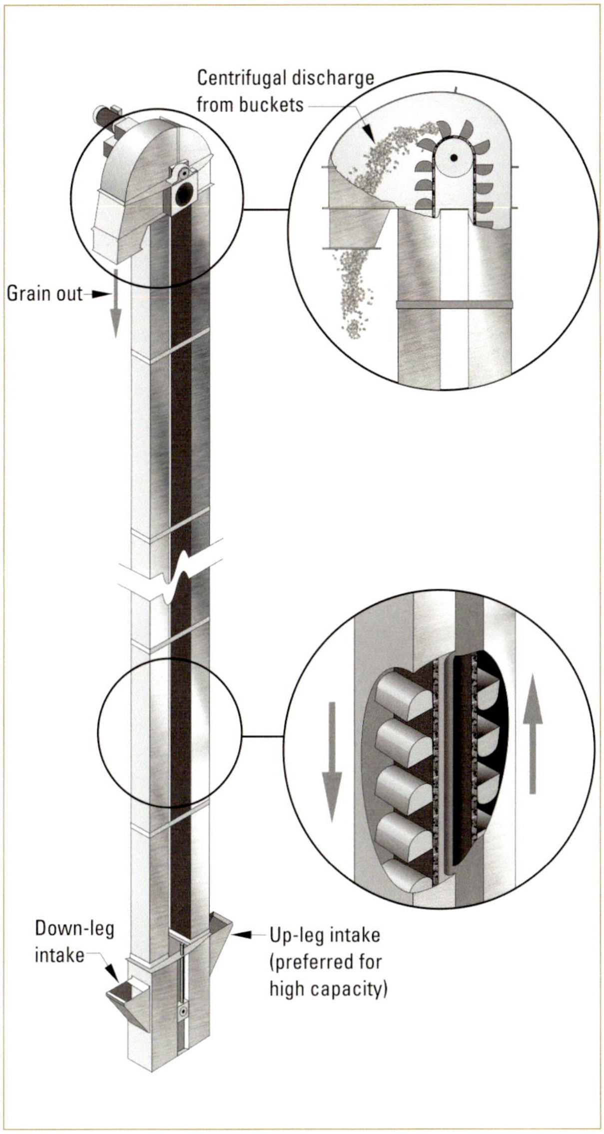


**FIGURE 3-24. Bucket elevator schematic.**

**TABLE 3-8. Typical capacity and power requirements for a bucket elevator.**

| Bucket size, in | Bucket spacing, in | Belt speed ft/min | Capacity bu/hr | Power requirements hp/10' height |
|---|---|---|---|---|
| 4x3 | 8 | 240 | 200 | 0.10 |
| | 6 | 270 | 300 | 0.125 |
| 6x4 | 4-1/4 | 270 | 550 | 0.20 |
| | 4-1/4 | 335 | 700 | 0.25 |
| 7x5 | 8 | 335 | 900 | 0.30 |
| | 6 | 335 | 1,200 | 0.33 |
| 9x5 | 7 | 265 | 1,600 | 0.5 |
| | 6 | 300 | 1,800 | 0.5 |
| 9x6 | 12 | 385 | 1,500 | 0.625 |
| | 6 | 385 | 3,000 | 1.25 |
| 12x7 | 10 | 565 | 5,000 | 2 |
| 15x7 | 9 | 565 | 7,500 | 3 |
| 14x8 | 10 | 650 | 10,000 | 4 |

**TABLE 3-9. Estimated capacity of downspouts for dry grain.**

For 45-degree slope and generally unrestricted flow. Based on an industry recommendation of 65 bu/hr/in$^2$ of downspout open area as a conservative estimate.

| Downspout dia., in | Capacity bu/hr |
|---|---|
| 4 | 800 |
| 6 | 1,800 |
| 8 | 3,300 |
| 10 | 5,100 |
| 12 | 7,400 |
| 14 | 10,000 |
| 16 | 13,100 |

### Capacities

The capacity of a bucket elevator can exceed 30,000 bu/hr. Table 3-8 shows common farm capacities ranging from 1,000 to 6,000 bu/hr. For a single-bucket elevator without a dump pit, at least 3,000 bu/hr is suggested. If the pit for receiving wet grain is about as big as the largest grain hauling vehicle, a lower capacity leg can work. Table 3-9 gives gravity spout capacities. Table 3-10 shows common applications for bucket elevators.

When buying a bucket elevator, consider that capacity can often be doubled with only a 20% cost increase. Bucket elevators are relatively easy and less expensive to build taller when erected, but are difficult and expensive to make taller later. Money spent on greater capacity and discharge height may help future expansion. Special consideration is required for drop heights over 80 feet or spout angles greater than 45 degrees. Because higher grain velocities may greatly increase impact damage to corn, soybeans, and other fragile grains such as rice, consider using horizontal conveyors or installing grain decelerators in the downspouts every 30 feet when the discharge height of a bucket elevator exceeds 80 to 90 feet.

### Required height for gravity spout discharge

Select the height of a bucket elevator with care. As shown in Figure 3-25, several heights may be quoted for the same bucket elevator. Effective elevating height determines power requirements. Gravity spouting discharge height is the distance from the bucket elevator bottom to the distributor discharge, valves, or perhaps to the grain cleaner (for example, input elevation to gravity spouts). If the elevator

| TABLE 3-10. Bucket elevator capacities and matching screw conveyor and gravity spout applications. | | | |
|---|---|---|---|
| **Capacity bu/hr** | **Matching screw conveyor and gravity spout** | **Applications** | **Comments** |
| Less than 1,500 | • 4-6" conveyor in dry grain<br>• 6" conveyor in wet grain<br>• 4-6" gravity spouts | • Small farm needs<br>• Well suited to feed processing<br>• May serve as wet or dry grain for conveyor or smaller batch and continuous flow dryers<br>• Seed handling | Capacity is too low for grain receiving or loadout. |
| 1,000-2,000 | • 6-8" conveyors in dry grain<br>• 8" conveyors in wet grain<br>• 6-8" gravity spouts | • Small to medium farms, cash-grain or livestock<br>• Can be the primary conveyor to fill small continuous-flow or batch dryers. | Lowest recommended capacity for grain receiving and loadout on small livestock or cash grain farms. |
| 2,000-3,500 | • 8-10" conveyors in dry grain<br>• 10-12" conveyors in wet grain<br>• 8-10" gravity spouts | • Medium to large farms.<br>• Wet grain receiving and grain loadout<br>• Load-unload for medium to large continuous-flow dryers.<br>• Primary conveying method for transferring grain in the system. | Recommended for medium to large volumes where a high degree of convenience, reliability, and performance is desired. |
| 3,500-5,000 | • 10-12" conveyors in dry grain<br>• 12-14" conveyors in wet grain<br>• 10-12" gravity spouts | • Large farms.<br>• Primary grain receiving and loadout conveyor.<br>• Load and unload large continuous-flow dryer and transfer grain in the system. | Recommended for large to very large farms with multiple handling needs for drying, marketing, and feeding and where high convenience, reliability, and performance are desired. |
| 5,000-6,500 | • 12-14" conveyors in dry grain<br>• 14-16" conveyors in wet grain<br>• 10-12" gravity spouts | • Large farms and medium elevators.<br>• Primary grain receiving and loadout conveyor.<br>• Load and unload large continuous-flow dryers and transfer grain in the system | Recommended for large to very large farms and medium elevators with multiple handling needs for drying, marketing, and feeding and where high convenience, reliability, and performance are desired, including fast loading of trucks to haul grain to elevator or ethanol plant at high frequency. |
| 6,500-8,000 | • 14-16" conveyors in dry grain<br>• 16" conveyors in wet grain<br>• 12" gravity spouts | • Very large farms to large elevators.<br>• Primary grain receiving and unload conveyor.<br>• Load and unload largest continuous-flow dryers and transfer grain in the system. | Recommended for very large farms to large elevators with multiple handling needs for drying, marketing, and feeding and where high convenience, reliability, and performance are desired, including fast loading of trucks to haul grain to elevator or ethanol plant at high frequency. |

bottom is in a pit or above grade level, account for this when determining the required discharge height.

### Controls for bucket elevators

If a bucket elevator is overloaded, it can plug and stall. If it stalls and the motor is not stopped, the belt drive pulley can burn through the belt, causing the belt and buckets to fall. Not only is there potential for fire and mechanical damage, but unplugging and repairing a bucket elevator with a broken belt is difficult. Use a motion detection sensor to protect against the belt stopping while the motor still operates.

A load indicator or control to prevent overloading the bucket elevator is desirable. An ammeter on the elevator's drive motor can indicate load, set off an alarm, or control a relay to stop the feeding conveyor. To prevent grain from backing up the spouts, a bin-level sensor can

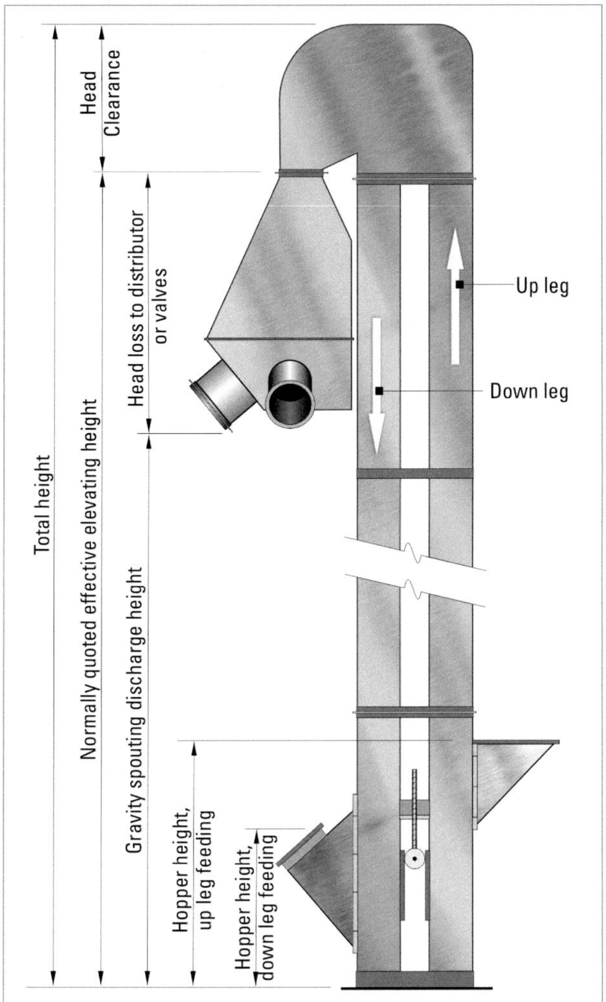


**FIGURE 3-25. Typically cited dimensions for bucket elevators.**
When determining required total height, consider that the bucket elevator may be in a pit, and/or the grade elevation may change from bucket elevator to storage.

activate an alarm and/or stop the bucket elevator motor when the bin is full.

Bucket elevators can be equipped with an anti-reverse mechanism or backstop device. This device prevents the drive wheel and belt from running backwards if the power or drive train fails when the buckets are loaded. Without this mechanism, grain in the loaded upside of the leg will fill the boot and lower portion of the leg making it very difficult to clean the grain out of the leg, boot and boot pit, and then restart the belt.

### Safety

Historical data show that many primary grain-dust explosions occur in bucket elevators. Using appropriate sensors such as monitors of bearing temperature, belt speed sensors and belt alignment sensors, and plug switches are highly encouraged. Dust explosion vents and dust explosion control devices should also be used to avoid grain dust explosions.

Overheated bearings in elevator boots with a buildup of grain dust and foreign matter around the bearings are primary causes of leg-related dust explosions. Bearing maintenance and housekeeping are the primary ways to minimize explosions related to grain dust in elevator legs.

Keep the shields on all drive assemblies for bucket elevators. Put guard rails on all service platforms, and put cages on ladders. Install resting platforms with rails on ladders per OSHA specifications. Have the guy wires positioned, installed, and tensioned by professionals. Bucket elevators supported by towers with stairs are safer and more convenient than bucket elevators and ladders supported by guy wires, but installation of towers with stairs costs more than using guy wires. Climbing to the top of a bucket elevator for servicing, maintenance, or curiosity is dangerous. Forbid climbing unless necessary. Require and use a safety harness with lanyards to secure workers to bucket elevator leg structures and ladders when doing high maintenance work.

## Pneumatic Conveyors

On-farm pneumatic conveyors are usually low-capacity conveyors that are PTO-driven or powered by electric motors. High-capacity, portable pneumatic conveyors powered by internal combustion engines are usually used for applications in commercial grain elevator facilities.

There are two types of pneumatic grain conveyors: positive pressure and negative pressure. Some units combine both types, that is suction/pressure units, by which grain can be vacuumed from a bin and pushed into a truck, as detailed in Figure 3-26.

Figure 3-27 illustrates how positive-pressure pneumatic conveyors have a positive displacement blower to push air and grain through the pipe. Grain flows into the air stream through an air lock.

Pneumatic conveyors can plug, so choose a unit that has easy access to clean out any part of the system. Clean the filters at regular intervals to prevent restricted airflow and reduced capacity. Negative-pressure conveyors can suck in a lot of fines and foreign material, especially when moving sunflower. These can plug the filter or build up on the blower vanes and stop the blower.

Positive-pressure pneumatic conveyors can be developed into a completely interconnected, push-button, grain conveying system. They adapt well to systems that require relatively low-capacity conveying over long distances (greater than 100 feet) to several different points and directions. Positive-pressure pneumatic conveyors are

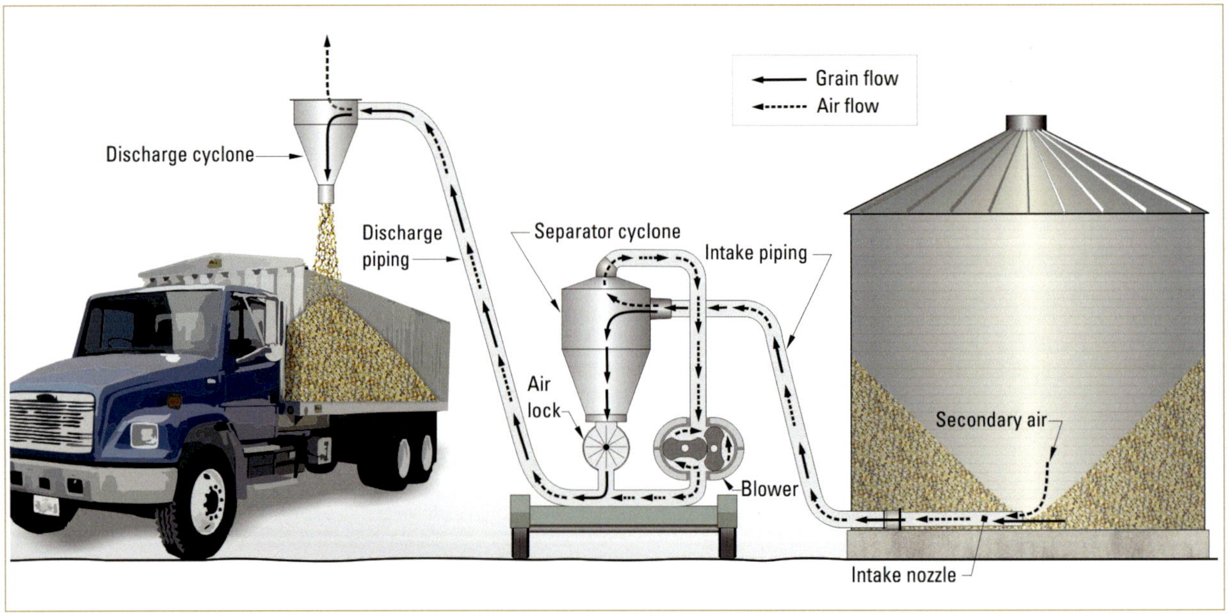


**FIGURE 3-26. Combination negative- and positive-pressure pneumatic conveyor.**
Typical conveyor for commercial grain elevator application. A blower pulls air and grain through the intake nozzle by negative pressure (suction) to the in-stream cyclone where they are separated. Air is pulled through a filter into the blower, compressed, and blown into the grain stream again, pushing grain to the discharge by positive pressure.

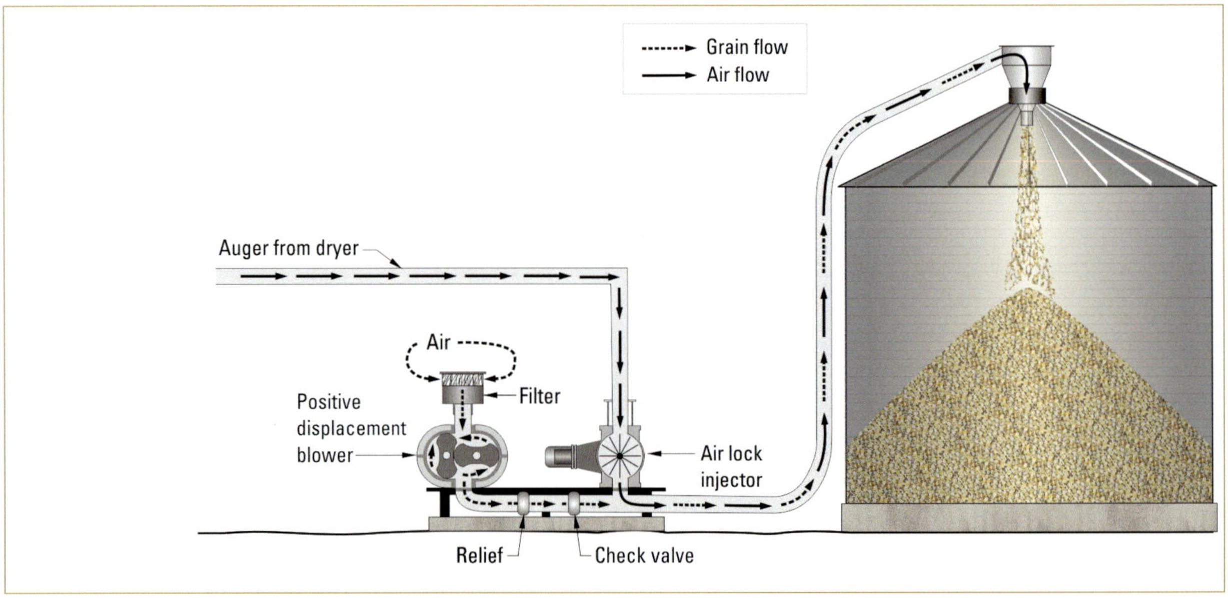


**FIGURE 3-27. Positive-pressure pneumatic conveyor.**
Typical on-farm pneumatic conveyor. Usually permanently installed and powered by an electric motor.

used mainly to deliver grain to storage from the dryer or receiving pit. Another primary conveying task is to unload storages, but the blower and injector unit must be moved to the bin for unloading, then back to a dryer or receiving area for refilling the bin.

Grain can be routed to alternate points by diverting grain flow in a Y-tube or by shifting a flexible tube to alternate lines at a manifold. The tubes can go above or below ground, up or down grade to follow terrain, and around corners to reach out-of-the-way bins.

Pneumatic conveying systems generally have a lower initial cost than bucket elevator systems, but have higher power requirements and lower capacity. Pneumatic conveying of wet grain at flowrates over 1,500 bu/hr (about the wet grain receiving rate of an 8-inch inclined auger) usually requires at least 75 horsepower. In contrast, a 2,500 to 3,000 bu/hr bucket elevator with an 80-foot discharge height can handle wet grain with only 12 to 15 horsepower. If a tractor is always available when needed, the high horsepower needs may not be a limiting factor. If another engine or electric motor must be purchased, then it is a large factor. Capacities over 1,000 to 1,500 bu/hr may not be realistic for electric motors. Pneumatic conveyors with flowrates over 1,000 to 1,500 bu/hr are usually PTO-driven.

A pneumatic conveyor is not as energy efficient as other conveying systems. Energy requirements are 2 to 2.5 kw-hr/100 bushel conveyed; a bucket elevator requires 0.5 to 0.75 kw-hr/100 bushel conveyed.

Pneumatic conveying works well for relatively low flowrates of dry grain, such as from a continuous-flow dryer or dryeration bin to storage, or from storage to a loadout bin. Flowrates of 200 to 500 bu/hr are usually adequate, requiring relatively small distribution tubes (4 to 5 inches), and power needs are easily met with electricity. Pneumatic conveyors can unload a batch dryer, but a surge bin is required to load the dryer quickly.

Tables 3-11 and 3-12 show the conveying rate of a given tube depending on grain type, tube length, number of turns (elbows), and change in elevation. Because capacity is generally listed for a short horizontal distance and a small vertical lift, capacity may be smaller than expected. Increasing conveying distance reduces capacity. Increasing lift reduces capacity considerably and will increase power requirement. Select the capacity of pneumatic conveyors based on required conveyor length and life from present use to future expansion.

One study on grain damage in pneumatic conveyors indicates that damage to corn in a 4-inch diameter system is not serious. Damage studies show grain impact can be a problem, but scuffing of grain as it slides along a tube surface is not, if pipe sections are properly aligned with smooth junctions.

Sharp turns can increase grain damage and will wear out the tube rapidly. Make all turns with a large radius. Blowers normally give higher air and grain velocities than do positive-displacement pumps, which can lead to higher grain damage. Research indicates that air velocity should not exceed about 5,000 feet per minute (ft/min) to minimize grain damage.

Airflow through grain inductors should be at least 3,500 ft/min pickup velocity to provide proper grain acceleration in the air stream. Straight, horizontal pipes downstream of a pressure-system grain inductor should be at least 8 to 10 feet long to allow grain to reach proper transit velocity along the pneumatic tube before any change in direction or elevation. Maintain the manufacturer's recommended air-to-grain volume ratio to minimize grain damage. Use a grain decelerator (usually a cyclone) at the discharge, as shown in Figures 3-26 and 3-27. Even though a pneumatic conveyor can easily reach a bin in any location, it is still

**TABLE 3-12. Typical pneumatic conveyor capacity and power.**

For moving dry grain 150' horizontally.

| Pipe dia | Capacity bu/hr | Power required hp |
|---|---|---|
| 3" | 400 | 10 |
| 4" | 700 | 15 |
| 5" | 1,200-1,500 | 20-30 |
| 6" | 2,000-2,500 | 40-50 |

**TABLE 3-11. Example pneumatic conveyor capacity.**

For a positive-pressure pneumatic conveyor with a 4-inch diameter line and a 20 hp blower with a 3/4 hp airlock.

| Vertical distance ft | Horizontal distance, ft | | | | | | | |
|---|---|---|---|---|---|---|---|---|
| | 25 | 50 | 75 | 100 | 125 | 150 | 175 | 200 |
| | - - - - - - - - bu/hr - - - - - - - - | | | | | | | |
| 20 | 915 | 876 | 833 | 793 | 755 | 718 | 682 | 649 |
| 40 | 858 | 821 | 787 | 745 | 709 | 675 | 642 | 610 |
| 60 | 805 | 772 | 735 | 700 | 667 | 635 | 604 | 574 |
| 80 | 758 | 726 | 692 | 660 | 628 | 598 | 569 | 541 |
| 100 | 714 | 684 | 653 | 622 | 593 | 564 | 537 | 510 |

*Source: Prairie Agricultural Machinery Institute Test Report No. 1860.*

important to locate bins and equipment in a well-planned layout for present use and future expansion.

Suction piping on suction/pressure units should be kept to a minimum as pressure conveying is more efficient than suction. Operators should find out the maximum suction and pressure operating limits of their pneumatic conveyor. If a unit is operated above the combined maximum limits, the rotary vane blower can quickly overheat, warping the blower. Once a blower overheats, the operating efficiency of the conveyor drops dramatically. The only solution to a warped blower is replacement, which is expensive. Check with your manufacturer to get proper operating guidelines.

CHAPTER 4

# GRAIN **DRYING**

*Klein Ileleji, Ph.D., Purdue University*
*Dirk E. Maier, Ph.D., P.E., Iowa State University*
*Kenneth Hellevang, Ph.D., P.E., North Dakota State University*

Grain harvested too wet for safe storage should be dried for maximum marketing potential and storage flexibility. Table 4-1 summarizes maximum harvest and safe storage moisture content recommendations for grains. Several pieces of equipment are important to the successful operation of a drying system. Read through this section completely to better understand the function and operation of each piece of equipment before designing your system for holding wet grain and drying and cooling grain.

**TABLE 4-1. Maximum moisture contents for grain harvest and safe storage recommended in the Midwest.**

Values are for good-quality, clean grain and aerated storage. Reduce safe storage moisture content by 1 percentage point for poor quality grain.

| | Maximum moisture content, %wb | | | |
|---|---|---|---|---|
| | | Storage period | | |
| **Grain type** | **At harvest** | **Up to 6 months**[a] | **6-12 months**[b] | **> 1 year**[b] |
| Shelled corn and grain sorghum | 30 | 15 | 13 | 13 |
| Soybeans | 18 | 13 | 12 | 11 |
| Wheat, barley[c] and oats | 20 | 14 | 13 | 12 |
| Flaxseed | 15 | 9 | 7 | 7 |
| Canola | 14 | 9 | 8 | 8 |
| Sunflower[d] | 17 | 10 | 8 | 8 |
| Edible beans | 16 | 16 | 13 | 13 |

[a]Up to 6 months from harvest refers to storage under winter conditions.
[b]6-12 months and >1 year storage refers to storage into the warm summer months.
[c]Malt barley should be stored at 12% moisture to maintain the germination.
[d]The maximum storage moisture is 8% for oil sunflower and 9% for confectionary sunflower stored at temperatures above 40 degrees.

*Goals for a well-designed drying system are the following:*

- Enable timely harvest of top quality grain.
- Provide safe and pleasant working conditions.
- Have the capacity to handle grain at the harvesting rate.
- Provide for easy expansion to accommodate future increased volume and harvesting rate.
- Have the capability to hold wet grain to be dried, dry wet grain, and cool dried grain efficiently.

## DRYING SYSTEM CAPACITY

Inadequate drying system capacity is often the bottleneck in a harvesting-drying-storage system. Provide enough capacity to receive and handle the daily harvest rate. Select equipment that allows for increasing capacity in the future.

Drying capacity depends on the entire drying system, not just the dryer. To determine the average receiving capacity for drying (bu/day), divide the annual grain volume (bu) to be dried by the number of operating days expected in the harvest season.

**For example:** There is a 90% chance that corn can be harvested in 20 to 25 working days out of the 45- to 50-day normal fall harvest season in the Corn Belt. However, in the Northern growing region, the harvest may need to be accomplished more quickly. Corn harvest starts in late October and needs to be completed by mid to late November since major snowfall can occur by mid-November.

To allow for harvesting other crops, tending livestock, and machinery breakdown, 15 to 20 corn harvesting days is a safer planning time for many farmers, especially with the current increased farming acreage. So, 150,000 bu/yr production (150 bu/ac from 1,000 acres) requires an average receiving capacity of 7,500 bu/day (150,000 bu/yr/20 days/yr). Additional wet-holding facilities may be needed to handle receiving grain from extended harvesting or higher harvesting rates.

# FANS

Selecting a fan is one of the most important considerations for grain wet-holding, drying, cooling, and aeration processes. The fan determines the amount of air that moves through the grain. The amount of air determines grain drying and cooling times. Fan operating characteristics are typically defined by static pressure and airflow.

The two fan types that are used to force air through grain are axial-flow, shown in Figure 4-1, and centrifugal, shown in Figure 4-2. These two categories are sub-divided into five fan groups: Tube-axial, Vane-axial, Low-speed centrifugal (1,750 rpm), High-speed centrifugal (3,500 rpm), and In-line centrifugal (3,500 rpm). Figure 4-3 displays typical axial and centrifugal fans.

Typically, axial-flow fans are noisiest but have a lower first cost than centrifugal fans. Axial-flow fans usually are selected for static pressure up to 3 to 4 inches. They gain heat from compression of air (approximately 0.75°F for each inch of static pressure) flowing across the motor housing, and they gain heat from "air-over" motors powering on fan wheels, which add 0.75 to 1°F.

Fan tip clearances for vane-axial fans should range from 1/16 to 1/8 inch for best performance. Wide fan tip clearances and/or off-center fan motors result in higher slippage (air backflow between fan blades and blower house) and lower static pressure. Consider mounting two vane-axial fans in series to develop static pressures comparable to those developed by centrifugal fans.

Centrifugal fans are more expensive than axial-flow fans but have a lower noise level. Centrifugal fans are usually selected for static pressures above 4 inches. For static pressures between 4 and 6 inches, compare individual fan performances and pick the fan that best fits your needs.

Generally, lower cost with higher noise levels is the trade-off when choosing a vane-axial fan over a centrifugal fan at similar airflow and static pressure conditions.

## Fan Noise

Fans can be considered as a point source sound generator. Fan noise can be measured using sound power meters, which measure sound on three scales: dBA, dBB and dBC scales. The dBA scale is the sound meter scale that reads closest to what the human ear hears. A reading difference of 3 dBA increases the sound that is heard by the human ear by approximately two times (twice as loud).

One sound measurement characteristic that is important in fan selection is determining whether the sound power reading drops by 3 dBA—or reduces by half each time the distance from the sound point source is doubled.

> **For example:** A vane-axial fan reading taken 5 feet directly in front of a vane-axial fan is 106 dBA. At 10 feet, the reading would be 103 dBA, or half as loud. At 20 feet, 100 dBA or 1/4 as loud; at 40 feet, 97 dBA or 1/8 as loud.

If the sound reading is taken at a 45-degree angle, the sound reading-noise level would be lower than it would be directly in front of the fan; at 90 degrees from the fan centerline, the sound would be even lower than at 45 degrees.

Sound levels from fans can be reduced by placing barriers such as walls or hay bales in front and around the fan to deflect the sound upward or absorb it. Building a sound box around the fan inlet with sound absorption materials lining the box will further contain or reduce the sound emissions from a fan. The sound box should not restrict air entering the fan or force the air to make bends as this will greatly reduce the amount of air moved by the fan.

## Fan Selection

Select a fan primarily for **airflow rate** and **static pressure**. Airflow is measured in cubic feet per minute (cfm), and static pressure in inches of water column (w.c.). Noise level may also be an important selection factor.

*The static pressure a fan operates against is affected by the following factors:*

- **Grain depth.** Resistance to airflow increases linearly with depth. Each added foot of grain adds about the same amount of airflow resistance.
- **Airflow through the grain, cubic feet per minute per bushel (cfm/bu).** For typical drying and aeration, doubling airflow triples airflow resistance and requires 4 to 6 times more fan horsepower depending on the fan. For example: 1 cfm/bu, 2" w.c., 3 hp vs. 2 cfm/bu; 6" w.c., 12 to 15 hp.

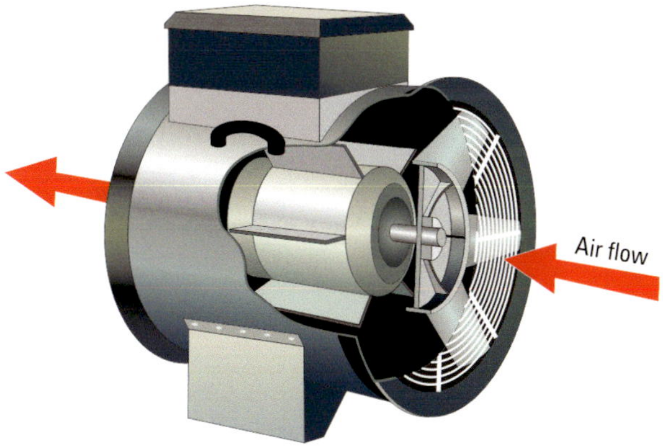


a. Detail

b. Example
*Photo courtesy of Sukup Manufacturing Co.*

**FIGURE 4-1. Typical axial-flow grain drying fan.**

a. Detail

b. Example
*Photo courtesy of Ken Hellevang.*

**FIGURE 4-2. Typical centrifugal grain drying fan.**

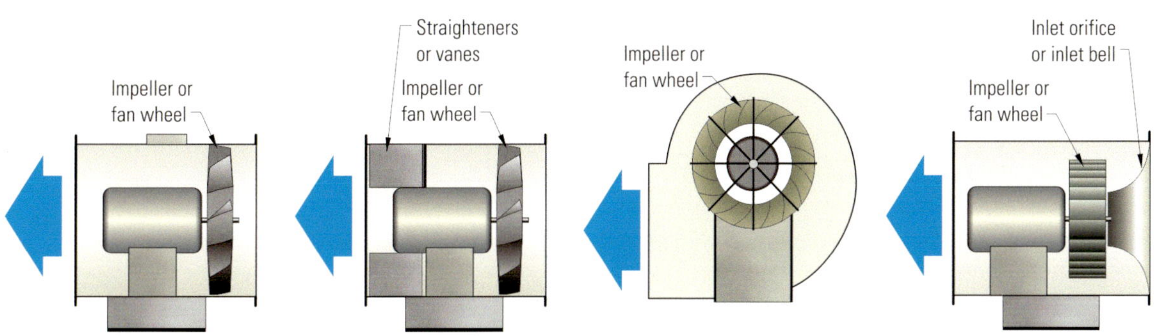


**FIGURE 4-3. Types of fans commonly used for grain drying, cooling, and aeration.**

- **Grain type.** Airflow resistance also depends on grain characteristics (for example, kernel size and shape). Resistance and power both increase when moving air through smaller-sized grains such as wheat or sorghum vs. corn or soybeans at the same grain depth.
- **Amount and distribution of fines and broken kernels.** Fines and broken kernels fill the void space between grain kernels and increase airflow resistance.
- **Grain compaction.** The extent of compaction depends on moisture content, filling method, and grain depth. Also, wind forces against bins and the length of storage time affect compaction. Vibration from vehicles on nearby roads and from trains passing by on nearby rail tracks also cause compaction.
- **Exhaust airflow area** (from headspace above grain) out of the bin with positive pressure fans and intake area (from headspace above grain) with negative pressure fans. Provide at least 1 $ft^2$ of exhaust or inlet area per 1,000 cfm.
- **Fan transition** and/or duct air velocities.

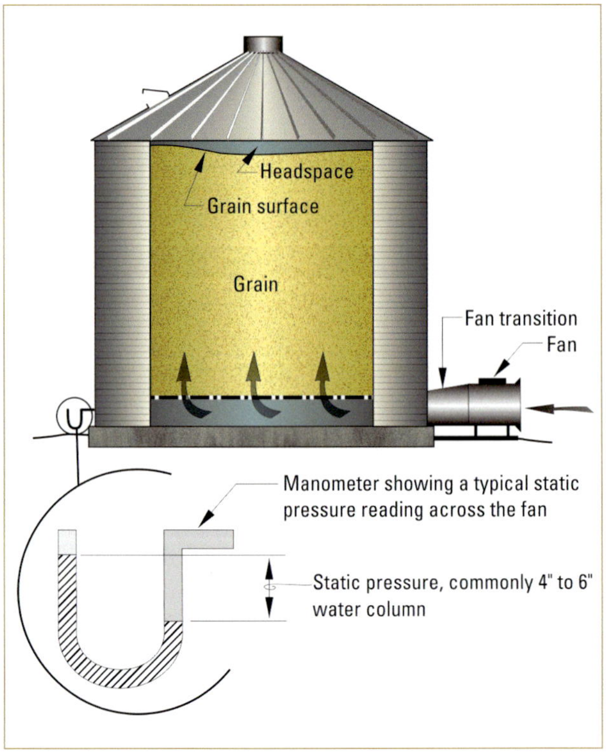


**FIGURE 4-4. Measuring static pressure in a bin.**

Figure 4-4 shows a static pressure measurement taken in the plenum of a bin. The manometer (U-tube), which senses the static pressure under the floor at the bottom of the bin, shows the static pressure at which the fan operates to overcome the resistance to airflow in the system.

A manometer can be a simple device made using a loop of 1/4- to 1/2-inch clear plastic tubing filled with water with a ruler or yardstick placed between the two tubing legs, mounted on the side of the bin. When the fan is started, the total distance between the two water levels, shown as 4 to 6 inches water column (w.c.) in the example in Figure 4-4, is the static pressure.

With a properly designed system, grain is the major source of resistance. The fan selected must deliver air at the static pressure needed to overcome the airflow resistances in the bin.

Each type and size of fan has unique performance characteristics, presented in a fan curve or fan table as airflow (cfm) vs. static pressure (inch of water). Table 4-2 and Figure 4-5 show examples of performance characteristics for the same two fans. These are just examples and should not be used in planning your system. Use the fan curve or table for your specific fan to evaluate its ability to provide the required airflow. The required fan curve or fan table is available from your fan dealer or manufacturer, as well as from the FANS online program (http://webapps.bbe.umn.edu/fans/). Use crop drying fans rated according to AMCA (Air Movement and Control Association) specifications, an independent fan-testing laboratory.

Fan tables from any reputable fan manufacturer include a fan's horsepower and airflow at various static pressures. The required fan motor horsepower is computed based on the static pressure (inches), total required airflow (cfm) and fan efficiency. Fan motor horsepower can be computed using Equation 4-1.

**EQUATION 4-1.**

$$EFMHP = (Q \times SP) / (6350 \times SE)$$

EFMHP = Estimated fan motor horsepower, hp
SP = Static pressure, inches
Q = Total required airflow, cfm
SE = Static efficiency, decimal. Ratio of the theoretical power needed to move the air against the static pressure to the actual power of the motor. Typical SE ranges from 0.40 for lower-horsepower fans to 0.70 for higher-horsepower fans.

**TABLE 4-2. Example of fan performance data. Values are airflow rate in CFM.**

Data presented is a composite from several manufacturers. Consult a comparable table or performance curve for the actual fans being compared. See the fan selection section of MWPS-29, *Dry Grain Aeration Systems Design Handbook*, Table 2.2.

| Power HP | Diameter in | Speed RPM | Static Pressure 1" | 2" | 3" | 4" | 5" | 6" | 7" | 8" | 9" | 10" |
|---|---|---|---|---|---|---|---|---|---|---|---|---|
| **Axial** | | | | | | | | | | | | |
| 0.5 | 12 | 3,450 | 1,500 | 630 | | | | | | | | |
| 0.75 | 12 | 3,450 | 1,700 | 750 | | | | | | | | |
| 1 | 14 | 3,450 | 2,880 | 1,050 | | | | | | | | |
| 1.5 | 12 | 3,450 | 2,200 | 950 | | | | | | | | |
| 1.5 | 14 | 3,450 | 2,800 | 1,300 | | | | | | | | |
| 1.5 | 16 | 3,450 | 3,500 | 2,400 | 1,300 | | | | | | | |
| 1.5 | 18 | 3,450 | 4,350 | 3,000 | 1,400 | | | | | | | |
| 3 | 18 | 3,500 | 5,700 | 4,600 | 2,650 | 1,400 | | | | | | |
| 5 | 24 | 3,500 | 10,500 | 9,000 | 7,000 | 4,600 | 2,900 | | | | | |
| 7.5 | 24 | 3,500 | 12,500 | 11,100 | 9,450 | 6,550 | 3,900 | | | | | |
| 10 | 26 | 3,500 | 15,500 | 14,00 | 12,250 | 9,500 | 5,800 | 3,400 | | | | |
| **Low-Speed Centrifugal** | | | | | | | | | | | | |
| 3 | | 1,750 | 4,580 | 4,230 | 3,820 | 3,350 | 2,550 | | | | | |
| 5 | | 1,750 | 7,800 | 7,000 | 6,250 | 5,550 | 4,600 | 3,300 | | | | |
| 7.5 | | 1,750 | 10,550 | 9,750 | 8,950 | 8,000 | 7,400 | 6,100 | | | | |
| 10 | | 1,750 | 13,3000 | 12,400 | 11,550 | 10,5000 | 9,550 | 8,500 | 7,300 | | | |
| **High-Speed Centrifugal** | | | | | | | | | | | | |
| 3 | | 3,500 | - | 2,950 | - | 2,550 | - | 2,120 | - | 1,650 | - | 1,000 |
| 5 | | 3,500 | - | 4,350 | - | 3,850 | - | 3,200 | - | 2,220 | - | 1,800 |
| 7.5 | | 3,500 | - | 5,700 | - | 5,100 | - | 4,500 | - | 3,800 | - | 2,900 |
| 10 | | 3,500 | - | 6,800 | - | 6,300 | - | 5,750 | - | 5,100 | - | 4,450 |
| **In-Line Centrifugal** | | | | | | | | | | | | |
| 3 | 18 | 3,500 | 3,800 | 3,600 | 3,400 | 3,000 | 2,500 | 1,900 | | | | |
| 5 | 24 | 3,500 | 5,500 | 5,000 | 4,400 | 4,100 | 3,900 | 3,600 | 2,800 | 1,800 | | |
| 7.5 | 28 | 3,500 | 6,200 | 6,000 | 5,700 | 5,500 | 5,200 | 4,800 | 4,500 | 4,000 | 3,500 | 3,000 |
| 10 | 28 | 3,500 | 7,700 | 7,300 | 6,800 | 6,500 | 6,300 | 6,000 | 5,400 | 5,100 | 4,800 | 4,400 |

*Source: EB-35 Natural Air/Low Temperature Crop Drying, (Hellevang, 1993).*

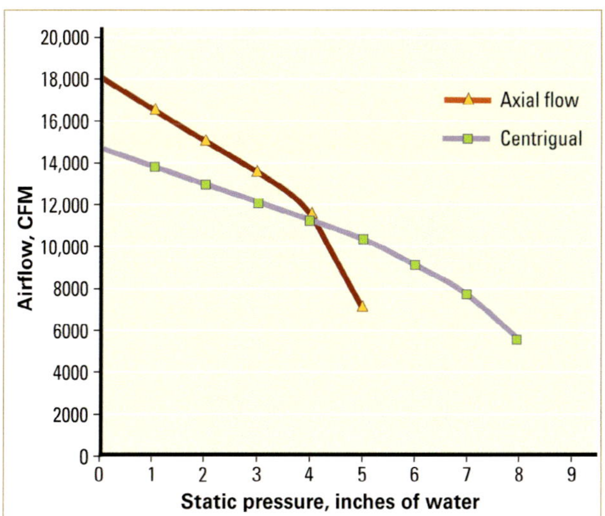


**FIGURE 4-5. Example fan curves.**
Each fan motor is rated at 10 hp. Example fan curves based on the average performance of 11 axial-flow fans and 7 centrifugal fans. These values are only for illustration; use the performance data or curves provided by the manufacturer for your fan. If that data is not available, avoid buying the fan.

*Fans can be compared using fan electrical efficiency, defined as cfm/watt. Fan operating costs are directly related to the cfm/watt of the fan.*

- Fan electrical efficiency typically varies widely between fans at the same static pressure.
- Fan electrical efficiency always reduces with the increase in static pressure; therefore, try to keep static pressure as low as possible.
- Fan electrical efficiency is not better for larger horsepower motors.

Select fans with the highest electrical efficiency at the expected operating static pressure. See MWPS-29, *Dry Grain Aeration Systems Design Handbook* for more details.

*Proper fan selection is based on moving the required amount of air through the grain. Many factors like those listed here affect the required airflow:*

- Is the fan for aeration of dry-grain in a storage bin?
- Is the fan for drying? If so, is it on a high-temperature or natural-air drying bin?
- Is the fan to be used for keeping wet grain cool?
- Is the fan to be used for dryeration or in-bin cooling?
- Is the fan to be used on more than one type of grain in the same bin?
- Is the fan on a continuous-flow, high-temperature stacked or tower grain dryer?

Fans should be selected using either a computer program or from calculations based on specific fan curves or charts. Use these powerful tools to assist you in making decisions on fan type, sizing, and selection. These technical aids are available from your extension agricultural engineer or knowledgeable commercial dealer. One of these tools is the FANS on-line program that can be accessed at http://webapps.bbe.umn.edu/fans/. This app allows the user to enter bin size (dia. in ft.), grain depth (ft), and crop type. It then calculates the fan horsepower requirements (hp) and allows the user to select fan size, type and arrangement (parallel or series). Once selected, airflow and static pressure for different depths of grain in the bin are displayed in table and chart format.

Airflow per bushel (cfm/bu) is the most useful characteristic for evaluating a fan because it directly relates to the time required to dry or cool grain. Determine airflow rate per bushel (cfm/bu) by dividing total airflow rate (cfm) by the number of bushels being dried, cooled, or aerated. For a given grain type, grain depth, and bin diameter, airflow rate (cfm/bu) can be used to compare different fans and fan arrangements in order to select the best option.

The rated horsepower of a motor is not a good way to compare fans. Fan performance is determined by its blade or impeller design. The horsepower required by a fan is determined by the airflow, static pressure, and fan efficiency. At different combinations of airflow (cfm) and static pressure (in. of water), the horsepower required for the fan varies. Table 4-3 shows the calculated fan horsepower for a vane axial-flow fan that is similar to the 10-horsepower fan shown in Table 4-2.

Table 4-3 illustrates the concept that the required fan horsepower varies considerably with operating conditions and static pressures. A maximum required horsepower input occurs at a specific combination of static pressure and airflow. The fan motor must be able to operate at the maximum horsepower on a continuous basis. A properly cooled motor can operate at a horsepower above the motor's rating. On the motor nameplate is a service factor, which indicates at what level of overload the motor can operate adequately. For example, a motor with a service factor of 1.1 can operate successfully at 10% above its nameplate rating. If the fan motor must operate at a horsepower above the motor's capability, the life of the motor will be reduced. The maximum calculated fan horsepower for the fan in Table 4-3 is 10.6, which is 6% above the rated horsepower. If the service factor for this fan is above 1.06, no problems should occur. Mounting the motor in the airstream (direct-drive, vane-axial fans—air-over motor applications) improves cooling for maximum motor performance. Problems can occur if the fan motor operates continuously at maximum conditions and is excessively loaded or improperly cooled.

All motors are rated and have a motor nameplate giving the electrical characteristics of the motor. Included on the motor nameplate is the full load current rating of the motor in amps. While a fan is operating, an electrician can measure the current to determine the load characteristics of the motor. If this is above the full-load current rating, the motor is overloaded.

Note that the motor will pull about 4 to 7 times its operating (locked rotor) amperage when starting depending on whether it is a single-phase or 3-phase motor. A single-phase motor has a higher starting multiplier than a 3-phase motor. Capacitor-start, single-phase motors generally have

**TABLE 4-3. Calculated fan horsepower at various airflows (cfm) for a motor rated at 10 horsepower.**

Fan horsepower depends on airflow, static pressure, and fan efficiency.

| | Static pressure (inches of water) | | | | | |
|---|---|---|---|---|---|---|
| | 0 | 1 | 2 | 3 | 4 | 5 |
| **cfm** | 18,000 | 16,500 | 15,000 | 13,500 | 11,500 | 7,000 |
| **fan hp** | 8.6 | 9.5 | 10.2 | 10.6 | 10.6 | 9.9 |

lower locked-rotor amperages than do single-phase motors without capacitor-assistance packages.

This fact is important because of the need to sequentially start multiple fans at a storage facility to prevent a huge current draw versus starting multiple fans at once. Preventing the starting of multiple fans at once can be achieved by using automatic sequencing control, typically an option in software that controls fan operations in a storage bin. **Huge penalties in the form of demand charges result in high electric bills due to high starting loads.**

The effect of corn depth on airflow and static pressure is shown in Table 4-4. As depth increases, static pressure increases, and airflow decreases, which reduces the drying and cooling capacity. Table 4-4 also shows the effect of adding up to 3 fans in parallel to a bin.

Table 4-4 is an example that illustrates principles using one, two and three 28-inch, 10-horsepower, inline-centrifugal fans and axial fans arranged in parallel configuration. See MWPS-29, *Dry Grain Aeration Systems Design Handbook* for more information about fan configurations. Values in Table 4-4 are based on a 30-foot diameter drying bin with shelled corn. The results of this example were generated using the FANS online program that can be accessed at http://webapps.bbe.umn.edu/fans/. The app does not describe specific equipment or recommendations for corrective action when incidents such as a fan stalling occurs.

**TABLE 4-4. Example performance of one, two, and three 28-inch, 10-horsepower fans installed in parallel on the same bin loaded with corn.**

| | | Axial flow | | | In-line Centrifugal | | |
|---|---|---|---|---|---|---|---|
| Number of fans | Corn depth ft | Static pressure in. water | Total airflow cfm | Airflow Rate cfm/bu | Static Pressure in. water | Total airflow cfm | Airflow rate cfm/bu |
| **One** | 2 | 0.90 | 19,500 | 17.24 | 0.17 | 6,115 | 5.41 |
| | 6 | 2.18 | 16,934 | 4.99 | 0.51 | 6,048 | 1.78 |
| | 10 | 3.04 | 15,005 | 2.65 | 0.84 | 5,982 | 1.06 |
| | 14 | 3.67 | 13,558 | 1.71 | 1.16 | 5,926 | 0.75 |
| | 18 | 4.13 | 12,368 | 1.22 | 1.48 | 5,878 | 0.58 |
| | 20 | 4.30 | 11,816 | 1.04 | 1.63 | 5,855 | 0.52 |
| | 25 | 4.65 | 10,671 | 0.75 | 2.02 | 5,797 | 0.41 |
| | 35 | 5.08 | 8,953 | 0.45 | 2.73 | 5,654 | 0.29 |
| | 45 | 5.28 | 7,666 | 0.30 | 3.42 | 5,537 | 0.22 |
| | 50 | 5.35 | 7,168 | 0.25 | 3.75 | 5,487 | 0.19 |
| **Two** | 2 | 2.14 | 34,058 | 30.11 | 0.45 | 12,122 | 10.72 |
| | 6 | 4.04 | 25,365 | 7.48 | 1.29 | 11,813 | 3.48 |
| | 10 | 4.80 | 20,346 | 3.60 | 2.08 | 11,566 | 2.05 |
| | 14 | 5.15 | 17,053 | 2.15 | 2.81 | 11,275 | 1.42 |
| | 18 | 5.32 | 14,723 | 1.45 | 3.51 | 11,046 | 1.09 |
| | 20 | 5.39 | 13,821 | 1.22 | 3.85 | 10,944 | 0.97 |
| | 25 | 5.53 | 12,051 | 0.85 | 4.63 | 10,648 | 0.75 |
| | 35 | 5.70 | 9,722 | 0.49 | 6.01 | 10,095 | 0.51 |
| | 45 | 5.82 | 8,229 | 0.32 | 7.22 | 9,613 | 0.38 |
| | 50 | 5.86 | 7,660 | 0.27 | 7.76 | 9,394 | 0.33 |
| **Three** | 2 | 3.21 | 43,885 | 38.80 | 0.80 | 17,973 | 15.89 |
| | 6 | 4.96 | 28,939 | 8.53 | 2.24 | 17,253 | 5.09 |
| | 10 | 5.34 | 21,823 | 3.86 | 3.52 | 16,565 | 2.93 |
| | 14 | 5.54 | 17,901 | 2.26 | 4.66 | 15,953 | 2.02 |
| | 18 | 5.66 | 15,354 | 1.51 | 5.66 | 15,352 | 1.51 |
| | 20 | 5.71 | 14,378 | 1.27 | 6.12 | 15,075 | 1.33 |
| | 25 | 5.81 | 12,475 | 0.88 | 7.19 | 14,438 | 1.02 |
| | 35 | 5.93 | 10,001 | 0.51 | 8.90 | 13,276 | 0.67 |
| | 45 | 6.01 | 8,431 | 0.33 | 10.29 | 12,338 | 0.48 |
| | 50 | 6.04 | 7,837 | 0.28 | 10.81 | 11,868 | 0.42 |

Consider grain depth when selecting fans because extra depth increases airflow resistance, which decreases drying rate. Drying capacity is directly proportional to airflow. For a given drying air temperature, any diameter bin, grain depth, or number of fans, the relative drying capacity can be determined by comparing total airflow.

**For example:** From Table 4-4, one 10-horsepower, axial-flow fan at a grain depth of 6 feet dries 1.37 times (16,934 cfm / 12,368 cfm) or 37% faster than at a grain depth of 18 feet. To compare different grain depths, bin diameters, and fan combinations, you must compare airflow rate (cfm/bu).

For a given bin, grain depth, and air temperature, adding an additional fan increases the airflow and drying rate. Compare airflow rates at a given depth in Table 4-4 to determine the effect of adding more fans.

**For example:** The relative drying capacity of two, 10-horsepower, axial-flow fans at a grain depth of 6 feet is 1.50 times (25,365 cfm/16,934 cfm) or 50% greater than one fan at a 6-foot grain depth. However, there is only a 19% increase with a corn depth of 18 feet using axial fans (14,723 cfm / 12,368 cfm), or 88% increase using two, in-line centrifugal fans (1,046 cfm / 5,878 cfm). Consider the cost and benefit of increased drying capacity when deciding to add another fan to a drying bin.

Keep in mind alternative crops that might be dried in the future. If you are drying large grains (corn or soybeans) initially but anticipate the need to dry small grains (wheat, sorghum, rice) in the future, vane-axial fans selected for large grains can be supplemented by mounting a second, identical vane-axial fan in series when higher static pressures are required. While this may save costs of buying one propeller-sized centrifugal fan to replace the axial fan, two axial fans in series may incur higher electrical operating costs.

## HEATERS

The temperature of the drying air affects drying capacity. Heat is added to the drying air using a fuel burner or electric heater with sufficient capacity to raise the temperature of the air and grain. Increasing drying air temperature increases the amount of water released by the grain that can be carried by the air and thereby increases drying capacity. The amount of heat energy needed from a burner/heater to raise the drying air temperature can be determined by using Equation 4-2.

**EQUATION 4-2.**

$$\text{Burner/heater capacity} = 1.1 \times AV \times (T_f - T_o)$$

Where, energy for the burner/heater capacity is measured in Btu/h; AV is the total airflow rate in cfm; $T_f$ is the final air temperature after the burner (°F), and $T_o$ is the initial air temperature before the burner (°F). Note that it takes one Btu to raise the temperature of 1 pound of water by 1°F.

**For example:** Heating air from 40°F to 220°F across the burner of a crossflow dryer with a total airflow rate of 30,000 cfm would require a burner/heater with a capacity of about (1.1 × 30,000 cfm × (220°F – 40°F)) = 5,940,000 Btu/h, or 5.94 million Btu/h. Increasing drying air temperature in a given dryer usually increases drying efficiency. As drying efficiency increases, more bushels of grain are dried per unit of heating energy.

LP (liquefied petroleum gas, or propane) is a common fuel for high-temperature dryers. To provide the necessary heat output, a vaporizer is required, especially in cold weather. Most burners have self-contained vaporizers. Natural gas (NG) is increasingly becoming the fuel of choice for grain dryers as supply lines are extended into rural areas. In most cases, it is a cheaper alternative to LP. However, access to NG and the fees incurred in making the connection to a NG line are the greatest impediment to NG use in on-farm drying systems. Farms that use NG do so because a gas line runs close to their property making access affordable. NG cost does not fluctuate as much as the cost of LP. Sometimes gas companies incentivize farmers to change from LP to NG. LP cost is typically expressed in dollars per gallon while NG is typically expressed in dollars per cubic foot or per therm. Therefore, in comparing LP with NG, it is best to use the same common unit of therms, which is the volume of NG that contains 100,000 BTU. One gallon of LP contains approximately 91,500 BTU. Thus, dividing 100,000 / 91,500 = 1.099, or about 1.1 gallons LP. Inversely, 91,500 / 100,000 = 0.915. Thus, one gallon of LP is equivalent to 0.915 therms of energy.

The cross-flow, high temperature grain dryer, with a burner capacity of 5.94 million BTU/hr, would use 5,940,000 / 100,000 = 59.4 therms of NG, or 1.1 × 59.4 = 65.34 gallons LP per hour. Using electric resistance heaters, (1.0 kWh = 3,412 BTU) the dryer would use 5,940,00 / 3,412 = 1,74 kWh.

It is advisable to consult your dryer dealer and local NG or LP supplier to discuss options before choosing which fuel to use.

As costs for traditional petroleum fuel increase, there is interest in alternative fuels for grain drying. Solar energy, crop residues (biomass) and biofuels have been used. They will likely be used in the future as interest in homegrown fuels and the desire to be energy independent and environmentally responsible increase. Biofuel sources of energy are biodiesel and biodiesel blends used to replace grain dryers fired by petroleum fuel oil. A grain dryer needs to be retrofitted with special burners using compatible accessories for methyl ester biofuels. Alternative fuel systems are not commonly used on farms in the U.S. and so will not be discussed further in this handbook.

## Condensation Problems

If the season's last batch is dried and stored in a high temperature bin dryer, a perforated wall liner or perforated tubes spaced 9 inches on center around the inner wall are needed to minimize sidewall moisture condensation that will occur during the cooling process. Perforated tubes or a wall liner should extend about 3/4 of the way up the wall height. Install perforated tube ends to start below the perforated floor and the wall liner no more than one foot above the perforated floor. If the last batch of the season is dried with low temperature air, properly cooled, and/or temporarily stored for marketing or feeding before spring, condensation protection is not needed.

Condensation can also occur in mechanical conveying equipment when moving hot grain from a high temperature dryer, particularly in colder weather. Condensation in a bucket elevator can accumulate in the boot and freeze. Check equipment frequently for condensation and wet trash. Condensation is generally not a problem in pneumatic conveying systems because the airflow can dry out the empty pipes.

Condensation also tends to form in downspouts and distributors and run into bins. Consider installing a manual-, gravity-, or spring-operated flapper valve, at the bin end of downspouts to keep warm, moist exhaust air from flowing up into spouts and conveyors. These valves stop warm air from going back into the downspout, preventing condensation from occurring and dripping back into the stored grain. The drawings in Figure 4-6 illustrate this concept. Another strategy is to install exhaust fans in the bin roof to keep the bin headspace under a slight negative air pressure to minimize moist exhaust air from moving up the downspouts. Size roof-exhaust fans to handle 25 to 50% more air than the capacity of the bin cooling fan when operated at a static pressure of at least 1/8 inch.

Condensation will likely occur and can be a problem in any bin filled with warm corn when outside temperatures are more than 20 degrees cooler than the corn. Unless the grain will be moved immediately after cooling, in-bin cooling of hot grain is discouraged when ambient temperatures are below 50°F.

## Moisture Testing

Collect dried grain samples from the entire grain stream as it is metered out of the dryer using a sampling device. Moisture testing freshly dried, hot grain requires special procedures because warm grain can cause the tester to give false readings. This can be corrected with an adjustment factor from the manufacturer of the moisture tester. If you cannot get warm or hot grain moisture calibration adjustment factors, you can determine an off-set for your moisture tester yourself. Put hot grain in a sealed, moisture-proof

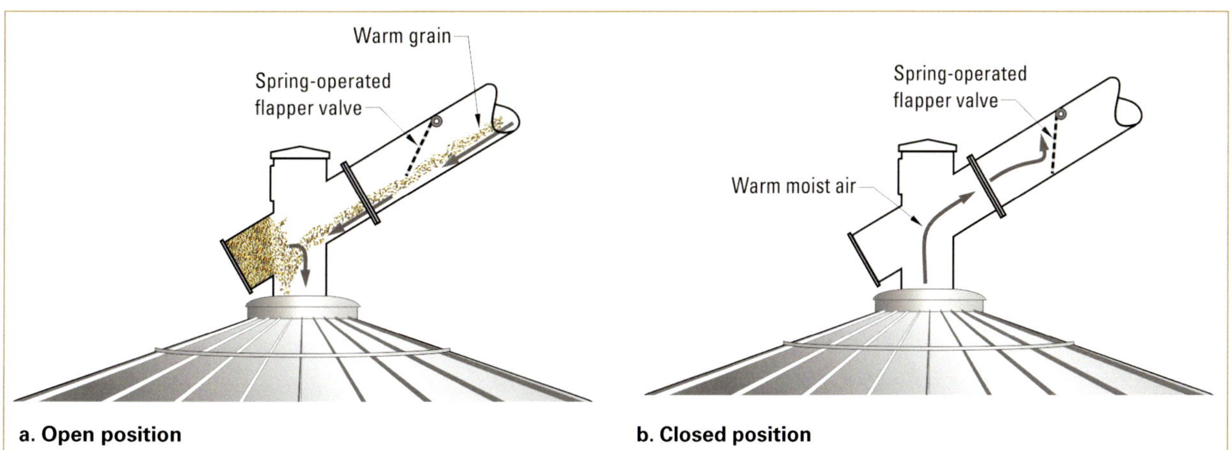


**FIGURE 4-6. Manually or spring-operated flapper valve.**

container like a quart-size freezer bag or a jar with a lid, and allow it to cool to room temperature before using your meter to measure grain moisture content.

Determine the off-set for your moisture meter by comparing the difference in moisture readings taken with your meter for samples cooled slowly and for hot samples tested immediately out of the dryer. The difference between the slowly cooled sample and the original hot sample reading is the off-set value for your meter. This value will vary for a particular moisture meter depending on the amount of moisture removed, the rate of moisture removal, removal range, and initial and final moisture content. Use these off-set values to correct future readings.

## GRAIN CONVEYORS AND GRAIN CLEANERS

A well-organized grain conveying system is needed to move grain between wet grain holding, the dryer, grain cooling, and storage. Conveyor size and speed is determined by process demands. Fast, high-capacity conveyors are typically used with batch-drying systems to allow for fast loading and unloading of batches. Slower, low-capacity conveyors are needed for continuous-flow drying systems.

Cleaning grain before drying improves airflow and drying capacity. Fines, foreign material, and broken kernels can accumulate on a perforated drying bin floor or in the void space between kernels, blocking airflow and reducing drying capacity. Fines and foreign materials also may accumulate on sensors and could cause them to give faulty readings. Fines and thrash accumulation in grain dryers have also been cited as the leading cause of grain dryer fires. See Chapter 3 of this handbook for more information about sizing and selecting grain conveyors and grain cleaners.

## HOLDING WET GRAIN

Without adequate wet-holding capacity, dryer capacity can limit harvest rate. The primary purpose of wet-holding is as a buffer so the dryer can operate through the night and over the weekend when harvest is stopped or slowed. This is especially needed with all high-temperature dryers so they can operate automatically and continuously for up to 24 hours per day. Effective ways to hold wet grain before drying range from a few hours in wagons, trucks, or hopper bins, to a month or more in a natural air or low temperature drying bin, as long as the allowable storage time is not exceeded. Table 5-2, page 95, displays maximum storage life for shelled corn. Wet grain is commonly held for several hours to a few days in hopper bins with fans to remove the heat of grain respiration. Figure 4-7 shows a typical wet-grain holding bin. Aeration is needed to minimize the heat of grain respiration that can damage wet, warm grain.

Multiple wet-grain holding bins allow for longer harvesting days, drying the rapid harvest of a custom operator, and allows harvest to continue if the dryer stops. Several days of drying capacity can accumulate in well-aerated, wet-holding bins for drying later during poor harvesting weather, while waiting for more grain to be ready for harvest, or while waiting for a custom operator to return.

A well-planned wet grain holding system greatly increases the flexibility and effectiveness of a harvest-to-storage system. Size wet holding capacity based on the dryer's maximum capacity and the maximum harvesting rate.

**EXAMPLE 4-1.**

Determine the required minimum wet-grain holding bin capacity and drying capacity to handle a harvesting rate of 1,250 bu/hr. Assume harvesting is conducted for 16 hours each day and all harvested grain is dried in a 24 hour period.

**SOLUTION 4-1.**

Daily harvesting capacity is:

$$1{,}250 \text{ bu/hr} \times 16 \text{ hr} = 20{,}000 \text{ bu}$$

Assume the dryer runs 24 hr/day; the minimum drying capacity is:

$$20{,}000 \text{ bu} \div 24 \text{ hr} = 833 \text{ bu/hr}$$

Minimum wet-holding capacity is the difference between the volume of grain harvested and dried during the harvesting period each day. Drying capacity during the 16 hours of harvesting is:

$$833 \text{ bu/hr} \times 16 \text{ hr} = 13{,}328 \text{ bu}$$

Minimum wet holding capacity is:

$$20{,}000 \text{ bu} - 13{,}328 \text{ bu} = 6{,}672 \text{ bu}$$

Plan for at least 50% more wet-holding capacity to handle longer harvesting times, carry-over from previous days, reduced drying capacity due to higher wet grain moisture content, or down time for repairs. For this example, select an operating, wet-grain holding capacity of 10,000 bu (6,672 × 1.5) preferably in two 5,000 bu hopper bins.

▶

This allows the dryer to keep up with harvesting and not create a bottleneck. All grain harvested each day will be dried in a 24-hour period. The wet holding bins will be empty and ready for the next day's harvest.

Grain respiration and fungi and bacteria activity cause wet grain to start spontaneous heating almost as soon as it is harvested. The heating rate depends on the type of grain, grain moisture content, temperature, and level and type of physical damage.

Aerate wet grain to prevent rapid temperature increases that can seriously damage the grain. Aerate wet grain continuously to keep it cool. Preventing blue eye mold is the main concern in wet corn. Aerated wet-grain holding bins require properly-sized fans and good air distribution. Provide wet holding bins with at least 0.5 cfm/bu (cubic feet of air per minute per bushel) to cool the full grain depth.

**For example:** A 10,000 bu wet-grain holding bin needs a fan that can deliver at least 5,000 cfm (10,000 bu × 1/2 cfm/bu). At 0.5 cfm/bu, air travel through the wet grain is about 1 minute. With good air distribution, 0.5 cfm/bu aeration is designed to completely cool a full hopper bin in about 24 to 30 hours. Given that wet grain is continuously drawn out of the hopper bin, the effective cooling time of the wet grain is faster. Corn harvested in northern regions at temperatures below 40 degrees may be adequately cooled with an airflow rate of 0.25 cfm/bu, cooling the entire bin in 48 to 60 hours.

Although perforated floors distribute air best, they are not feasible in hopper-bottom bins. Properly designed half-round duct systems placed down the hopper slope with the duct lower end open works well in wet holding bins. Air exhausting from the half-round duct into the hopper will spread uniformly as it flows upward through the grain bulk. Ducts need enough perforated surface area to provide an air velocity from the half-round duct into the grain bulk at about 35 feet per minute.

**For example:** In a 10,000-bushel bin with 0.5 cfm/bu, or 5,000 cfm, the surface area needed would be: 5,000/35 = 143 square feet. Round perforated ducts are considered to have only 75% of the perforated exhaust area. For round ducts in this example, divide the 143 square feet by 0.75; thus, round ducts should provide a total of 143/0.75 = 191 square feet. The open end of the duct offsets part of the 75% reduction factor. For large hopper bins, two to four ducts with individual small fans may be the most practical way to get the duct area needed for this high aeration rate.

**FIGURE 4-7. Wet grain holding bin system.**
It is important to have aeration fans running in wet-holding bins to minimize the chance of heating from grain respiration.
*Photo courtesy of Dirk Maier.*

Extension agricultural engineers or knowledgeable commercial representatives can help you select the correct fan and air distribution system for wet-grain holding bins.

Wet holding time before drying can greatly reduce the grain's overall storage life. Table 5-2 relates moisture content and temperature to overall storage life for shelled corn. Table 5-2 is also valid for grain sorghum. There is limited storability data for other grains, but the same general principles apply.

Completely empty unaerated, wet-holding bins every day to reduce self-heating and deterioration. Empty aerated, wet-holding bins so that no more than 1/2 of the maximum storage life is used up in wet holding. This can be managed more easily with two or more wet holding bins which avoids disrupting receiving.

**For example:** According to Table 5-2, the maximum storage time for 24% m.c. corn at 55°F grain temperature is 17 days. The allowable holding time in an aerated, wet-holding bin is half of the maximum storage time, which is 8.5 days (17 / 2). Therefore, a holding bin with 24% m.c. corn should be emptied at least every 8.5 days.

# DRYING SYSTEMS

When considering what type of drying system to install, consider the two primary alternatives: bin dryers or self-contained dryers.

## Natural-Air/Low-Temperature (NA/LT) Bin Dryers

In this handbook, natural air/low temperature drying means slow drying with natural air or air heated up to 10°F above ambient (NA/LT). Figure 4-8 illustrates a typical NA/LT bin dryer. Low temperature dryers require little grain handling, which is a big advantage of this system. The entire drying system (wet holding, drying and storage) is contained in the bin, so the process is called in-bin drying. With today's large farm sizes (1,000 acres and more), farmers who use in-bin drying systems would need to utilize several bins in order to dry their crop. A NA/LT bin dryer has a full perforated floor, high-capacity fan(s), a grain spreader, grain unloading equipment, roof exhaust vents, and eave opening that provide 750 to 1,000 cfm/sqft of roof vent area, and can also include power-exhaust roof fans and in-duct heater(s).

Over the years, research has been conducted on low-temperature drying for corn, sorghum, wheat, soybean, canola, barley, sunflower, rice, and other grains. Models have been developed to predict moisture removal for NA/LT in-bin drying systems. Low-temperature drying can be adapted to most grains, but the discussion in this handbook relates directly to corn, which is one of the predominant crops grown in the Midwest. Contact your state Extension agricultural engineer for assistance with using low- temperature drying with other grains, or refer to Extension bulletins that address one or more of the crops listed.

Adequate airflow (cfm/bu) is the key to successful low-temperature drying. As discussed in the fan section, drying capacity is directly related to airflow. Adding heat does not reduce the airflow needed to safely dry, but may cause over drying. Increasing the airflow dries grain faster and reduces the chance of spoilage, but greatly increases the fan horsepower required.

Figure 4-9 gives recommended minimum airflow rates (cfm/bu) for selecting fans for NA/LT corn drying in the North Central region of the U.S. Increasing airflow rates (cfm/bu) requires larger fans with more power and higher initial and operating costs. Airflow rates need to be selected that will safely dry the grain before deterioration occurs.

Select a fan to provide at least the airflow rate for your area when the bin is full. These airflow rates were developed by engineers from around the Midwest in part using computer simulation models and extensive historic weather data set, and were published in several state Extension bulletins. Contact the Extension agricultural engineer in your state for more specific values for your region.

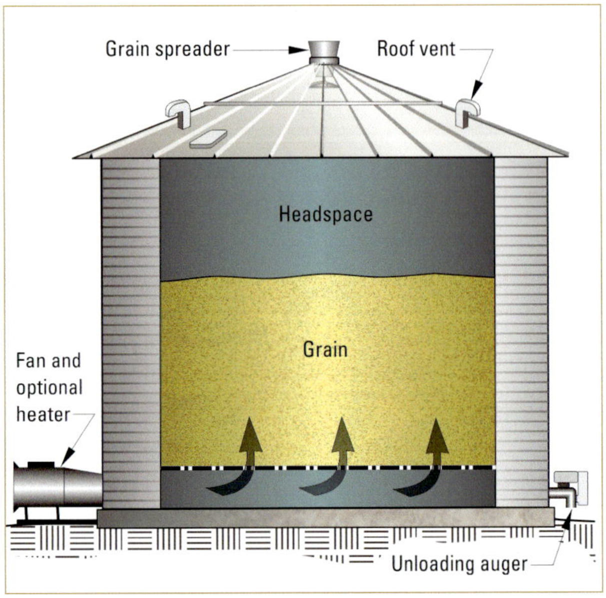


**FIGURE 4-8. Typical natural-air/low-temperature bin dryer.**

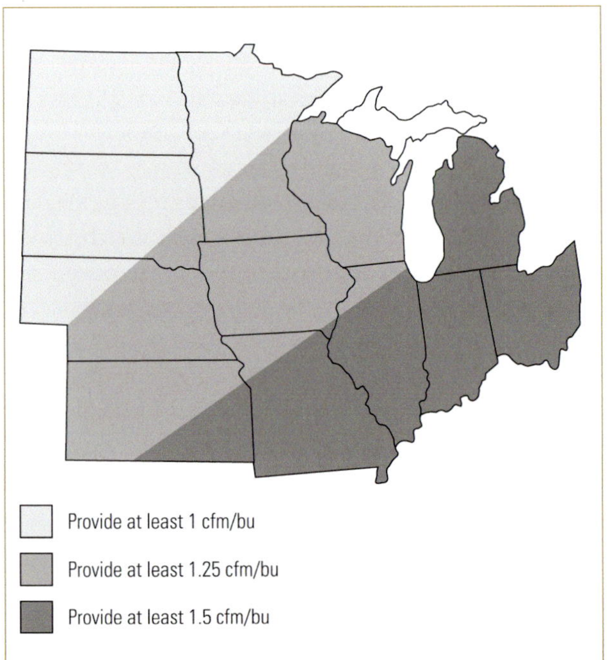


**FIGURE 4-9. Recommended minimum airflow rates for natural air/low-temperature drying of corn at 21% moisture content or less.**

*There are three basic procedures for filling a natural air/low temperature drying bin:*

1. **Single filling.** Load the bin as rapidly as harvest permits in one or two days.
2. **Layer filling.** Load the bin with a specific amount to 1/4, 1/3, or 1/2 of bin depth each week.
3. **Controlled filling.** Similar to layer filling but the amount of grain added each time depends on grain moisture content and drying progress. Use of moisture cables can help guide decision making. It is best to be conservative (use shallower fill layers), especially when using the system for the first year or two.

All three procedures work if adapted to your specific grain harvesting situation. It is important to thoroughly understand the relationships between airflow, grain and air temperature, grain moisture content, and the grain depth that can be dried without grain spoilage using a NA/LT bin dryer. For more information, refer to Extension bulletins from your State Cooperative Extension service, or the on-line resources of this handbook.

The maximum moisture content for the single-fill NA/LT drying method with the recommended airflow is about 21% for corn, 15% for sunflower, 15% for flaxseed, and 17% for small grains. At higher moisture contents, it is advised to use layer or controlled filling.

NA/LT dryers successfully dry most grains to safe storage moisture content in the Western and West Central Corn Belt without adding heat. These areas may experience possibly 2 wet years out of 10, when the relative humidity during the harvest season is 80% or above for a week or more. When this occurs, adding supplemental heat up to about 10°F helps complete drying in the fall.

The Eastern Corn Belt and some Northern regions have relatively high humidity in the fall, possibly 5 or more years out of 10. High humidity in the fall results in poor field dry down. Incoming corn moisture contents may be 25% or more. Heat the drying air up to 10°F above ambient under these conditions, and use layer or controlled filling, or the drying rate will be very slow, and spoilage may occur. If corn dries in the field to 20% to 22%, air temperatures are below 50°F, and grain moisture content in the bin is at least as low as the incoming grain, then the filling procedure may be changed to a single-fill, full-bin method. However, if drying is not complete by mid to late November when air temperatures drop, the only alternative is to cool the grain below 35°F during the winter and complete drying the following spring.

To finish drying before winter, harvest grain that matures in the fall as early as possible to take advantage of good fall drying weather. However, warm weather can require adjustments in low-temperature bin-dryer operation. As air temperature increases, the spoilage rate of grain increases faster than drying rate. So, to dry grain safely with the layer- or controlled-filling methods, shallower layers are required during September and perhaps early October. With the single-filling method, wetter grain can be loaded in the bin in October rather than September. Carefully monitor and check the progression of the drying front in NA/LT drying bins while drying grain ideally with moisture cables that are now commercially available.

Drying small grains in June or July with a NA/LT dryer can require several adjustments in the filling rate, especially in regions with high humidity, such as the mid-south rice belt. Warm weather with high humidity is ideal for mold growth. Under these conditions, use layer filling with small grains over 17% moisture content after the total accumulated depth reaches 4 feet.

If heat is used with low-temperature drying, common heat sources are electricity, or LP gas. However, unneeded heat increases drying cost and over dries the bottom grain, causing undesirable weight loss. A drying strategy to avoid over drying of the bottom layer using weather data as feedback to the burner, called self-adapting variable heat (SAVH) was developed at Purdue University. This system has been adopted by several fan-burner control suppliers which are applying this, smart-control algorithm to save energy, reduce shrink loss, and preserve grain quality. The use of solar heat, also a potential source of supplemental heat for NA/LT in-bin drying systems, will not be discussed in this handbook.

NA/LT drying bins can include stirring equipment for controlling final grain moisture content and increasing airflow rate. Disadvantages are increased investment and maintenance costs and about 1.5 feet of storage depth lost to the stirring equipment drive mechanism mounted in the bin headspace and increased grain depth from the looser fill (lower bulk density grain) which occurs with stirring. Also, note that stirring should not take the place of best management practices such as grain cleaning and leveling during filling.

Stirring does not significantly increase corn damage. However, fines can sift to the drying floor with stirring, where they fill the void space in the bottom foot of grain. This will reduce airflow.

Grain cleaning before drying is recommended for all low-temperature drying and is essential for stir drying to minimize the accumulation of fines on the drying floor. Proper stirring reduces bulk density of the grain, which increases airflow up to 30%. The additional cost and storage loss incurred due to stirring equipment in NA/LT drying bins can be justified by the advantages of increased airflow and drying capacity in the bin, and by enabling uniform moisture content from top to bottom of the grain mass. However, moisture cables cannot be used in the same bin with stirring equipment.

Adequate airflow is essential for NA/LT drying. Shallower grain depths allow for higher airflow, faster drying, less overdried grain, and lower operating costs.

## High-Temperature Bin Dryers

High-temperature bin dryers use heated air and higher airflow rates to dry grain in bin-batch or continuous-flow processes. Two types of bin-batch dryers are on-floor and roof batch dryers.

High-temperature bin dryers commonly dry with air heated to between 110°F and 200°F. The grain's nutritional value is affected if grain kernels reach the highest drying air temperatures. If delayed cooling is not used, the higher temperatures in this range can affect physical grain quality; that is, more stress cracks and kernel brittleness, less color, and reduced germination may occur. Temperatures above 110°F for malting barley or 120°F for wheat with a moisture content of 15 to 16% or less can affect marketability. Keep drying air below 110°F for seed grain, food corn, and soybeans; and below 100°F for edible beans.

Drying and storage occurs usually in separate bins. The drying bin has a full perforated floor, drying fan, heater, grain spreader, power exhaust on the bin roof in large bins, and grain unloading equipment. Batch-drying bins may also have grain stirring equipment. One drying bin can dry grain for one or more storage bins.

High-temperature bin dryers are primarily for drying grain, but can be used to store the last batch of grain dried. Some storage problems can occur with storing grain in a drying bin. Conveying grain to storage can reduce storage problems by mixing drier and wetter grain. If the last batch is stored in the drying bin, mixing does not occur. To reduce storage problems, consider the following procedures:

Select shorter-sidewall (shallow) drying bins to minimize the potential for moisture differences between the bottom and top layers of grain. The shallow depth enables the drying front to reach the topmost layer at a faster rate than would occur with a deeper bin.

Install a perforated wall liner or perforated tubes around the inner wall to minimize sidewall moisture condensation. For installation details, see the discussion of condensation problems later in this chapter.

After the last batch has been dried, run the drying fan extra hours with unheated air to cool the grain and dry out condensation on the walls and underside of the roof.

If grain moisture content is less than 18%, cool the grain and maintain the temperature between 35°F and 40°F, and market or feed the grain before spring.

### On-floor bin-batch dryer

An on-floor bin-batch dryer dries one layer of grain on the perforated floor at a time. Limit batch drying depth so the moisture content of the dried batch varies less than about 5 percentage points (for example, from 16 to 11%) between top and bottom. Moving the dried batch to storage mixes the wetter top grain with the drier bottom grain for safer storage. Aeration reduces the moisture differences over time. As grain depth increases, drying capacity decreases; grain moisture spread between bottom and top increases, and managing the drying process becomes more difficult. Large-capacity bins require high static pressures to move air through the grain mass, and achieving uniform drying across the bulk can be a challenge. The use of vertical stirring equipment is helpful to reduce the bulk density of grain and facilitate drying. Care should be taken to operate stirring equipment according to the manufacturer's filling recommendations and crop type.

High-temperature bin dryers can be equipped with or without stirring equipment. Recommended grain depth for drying is 2-1/2 to 4 feet without stirring equipment, shown in Figure 4-10, and 6 to 9 feet with stirring equipment, shown in Figure 4-11. Proper management practices are necessary for successful drying at greater depths, even with stirring systems. Stirring equipment allows for greater depths and assures more uniform moisture content throughout the grain mass. However, stirring equipment is available only to a certain depth (maximum 24-foot), and depth of equipment is based on the type of grain. Table 4-5 shows batch sizes for several bin diameters.

With an on-floor bin-batch dryer, grain is usually cooled immediately in the dryer to minimize the time grain is in the bin. An alternate, delayed cooling method with high-temperature bin-batch dryers is to turn the fan off after drying for at least 4 hours to allow the grain to steep (kernel moisture moves from the kernel's center to the kernel's outer layers and kernels of different moisture content within the batch equilibrate to similar moisture contents) before cooling. Then turn the fan on again to finish drying the grain by evaporative cooling. See the section on cooling later in this chapter for more details about this process.

Figure 4-12 shows one of several effective bin arrangements. For bins without stirring equipment, select fan capacity, drying air temperature, and grain depth so at least one batch can be harvested, dried, cooled, and moved to storage each day. By adding a wet-grain holding bin and fast handling equipment, it may be possible to dry two or

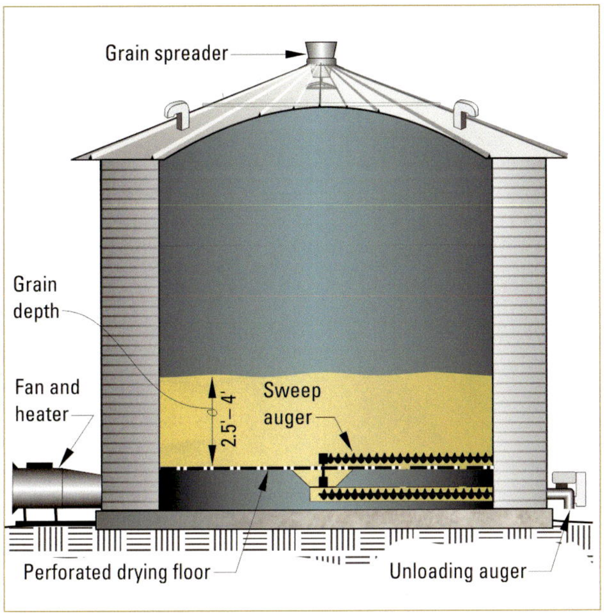


**FIGURE 4-10. Batch-bin dryer without stirring device.**

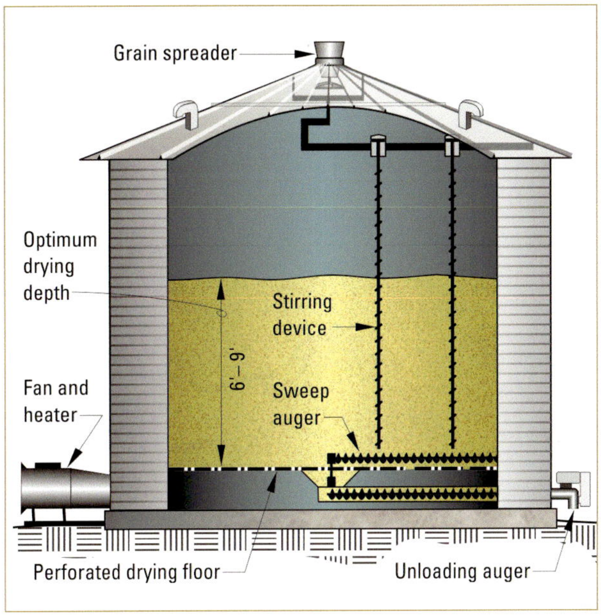


**FIGURE 4-11. Batch-bin dryer with stirring device.**

three batches per day, depending on grain type and moisture levels.

If drying only one batch a day, you can load wet grain directly into the drying bin. For more than one batch per day, hold wet grain in wagons or trucks, or add a holding bin. Transferring grain from wet holding to the bin dryer can be controlled automatically. Batch drying requires a high-capacity conveyor to minimize loading and unloading time and maximize drying capacity. Low capacity, inconvenient grain handling equipment slows bin-batch drying. Regardless of batch size and drying and cooling procedures, provide minimum conveying capacity of at least 2,000 bu/hr; 3,000 to 4,000 bu/hr is preferred to unload a 1,000 bu trailer in 15 to 20 minutes. A permanent, gear-driven sweep auger eliminates moving a motor-driven sweep auger in and out of the bin to unload each batch and eliminates hazards created by working in confined spaces. Another possibility is to install an under-floor auger with

**TABLE 4-5. Bushels per batch in bin dryers.**

| Bin dia. ft | Bin area ft$^2$ | bu/ft of depth | Approximate batch size, bu | | | | | | |
|---|---|---|---|---|---|---|---|---|---|
| | | | Grain depth, ft | | | | | | |
| | | 1 | 2.5 | 4 | 6 | 9 | 12 | 15 | 18 |
| 18 | 254 | 204 | 509 | 814 | 1,221 | 1,832 | 2,443 | 3,054 | 3,664 |
| 21 | 346 | 277 | 693 | 1,108 | 1,663 | 2,494 | 3,325 | 4,156 | 4,988 |
| 24 | 452 | 362 | 905 | 1,448 | 2,171 | 3,257 | 4,343 | 5,429 | 6,514 |
| 27 | 573 | 458 | 1,145 | 1,832 | 2,748 | 4,122 | 5,497 | 6,871 | 8,245 |
| 30 | 707 | 565 | 1,414 | 2,262 | 3,393 | 5,089 | 6,786 | 8,482 | 10,179 |
| 33 | 855 | 684 | 1,711 | 2,737 | 4,105 | 6,158 | 8,211 | 10,264 | 12,316 |
| 36 | 1,018 | 814 | 2,036 | 3,257 | 4,866 | 7,329 | 9,772 | 12,215 | 14,657 |
| 39 | 1,195 | 956 | 2,389 | 3,823 | 5,734 | 8,601 | 11,468 | 14,335 | 17,202 |
| 42 | 1,385 | 1,108 | 2,771 | 4,433 | 6,650 | 9,975 | 13,300 | 16,625 | 19,950 |
| 45 | 1,590 | 1,272 | 3,181 | 5,089 | 7.634 | 11,451 | 15,268 | 19,085 | 22,902 |
| 48 | 1,810 | 1,448 | 3,619 | 5,791 | 8,686 | 13,029 | 17,372 | 21,715 | 26,058 |
| 51 | 2,043 | 1,634 | 4,086 | 6,537 | 9,806 | 14,708 | 19,611 | 24,514 | 29,417 |

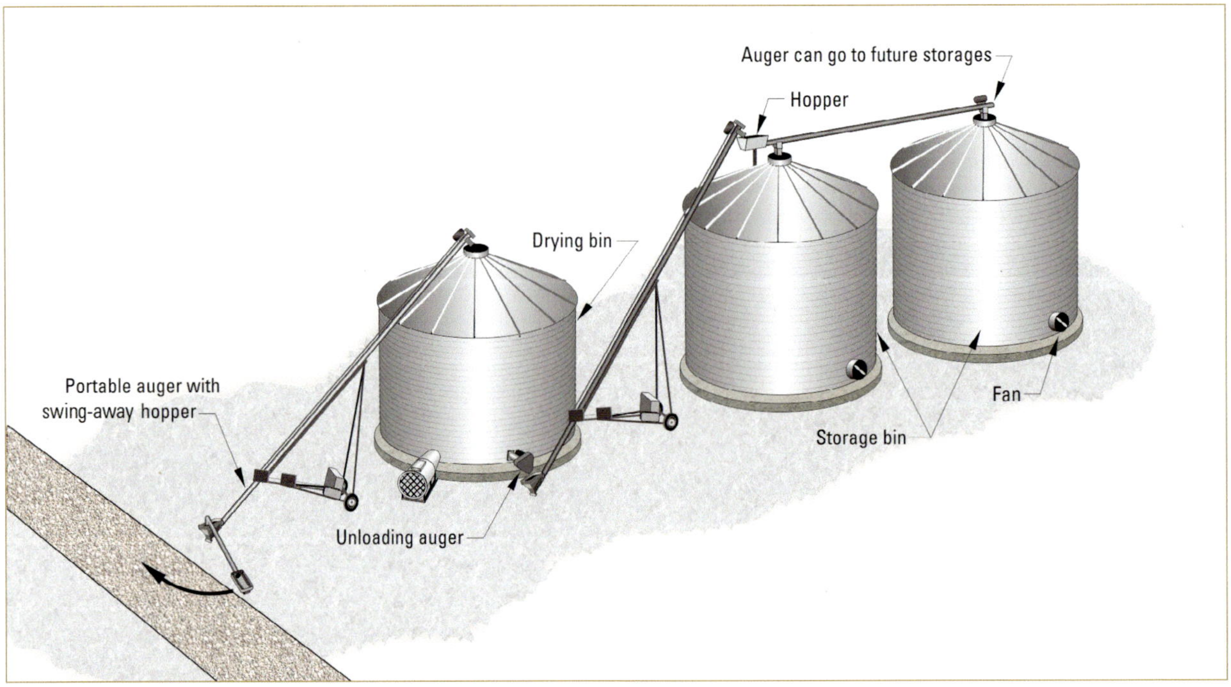


**FIGURE 4-12. High-temperature bin-batch drying system.**
Load the dryer using a portable auger. Another portable auger conveys dried and cooled grain to the overhead auger and storage bins.

multiple inlets, so a motor-driven sweep stays in place between batches. Place the sweep directly over the multiple inlets before loading the batch. Grain flowing into the inlets reduces grain depth over the sweep and permits it to be started. If stirring equipment is used, vertical stirring augers must clear sweep augers and motor by at least 2 to 4 inches.

Sweep augers must have back shields behind the auger to remove grain and minimize the buildup of fines on the perforated floor. Accumulated fines reduce airflow and drying capacity and can cause bin fires. They also are spots for insect pest accumulation and hotspot development if the grain is kept in the bin rather than moved to another bin. After every three or four batches, remove fines from the perforated floor to assure proper air movement and reduce fire risk.

Compared with other drying systems, bin-batch dryers often cost less per unit of drying capacity, but they require much more labor. Someone usually needs to supervise the transfer of each batch to storage, filling the next batch, and leveling grain in case spreaders do not operate properly. Consider a 24-hour time clock for convenient control of a bin-batch dryer. Set the timer to turn the burner off when the batch is estimated to be dry. The fan continues to run and cool the grain for subsequent transfer to storage.

### Roof bin-batch dryer

Roof bin-batch dryers, like the one shown in Figure 4-13, are in-bin dryers with drying floors suspended in the headspace of the bin. They are also known by some manufacturers as roof-top or top-dry grain drying systems. Roof bin-batch dryers dry a layer of grain about 2.5 feet deep on a cone-shaped, perforated floor mounted under the bin roof. New roof-bin dryers can be operated in either batch or continuous-flow modes. Capacities (bu/h) for these two modes differ. The approximate batch size for different diameter bins is shown in Table 4-6. However, the drying capacity in bushels per hour depends on the percentage point removal of moisture from the grain. Wet grain is loaded onto the coned, perforated floor for drying.

**TABLE 4-6 Approximate batch capacities of roof bin-batch dryers.**

| Bin dia ft | Batch size bu |
|---|---|
| 18 | 540 |
| 21 | 750 |
| 24 | 1,000 |
| 27 | 1,250 |
| 30 | 1,500 |
| 36 | 2,100 |

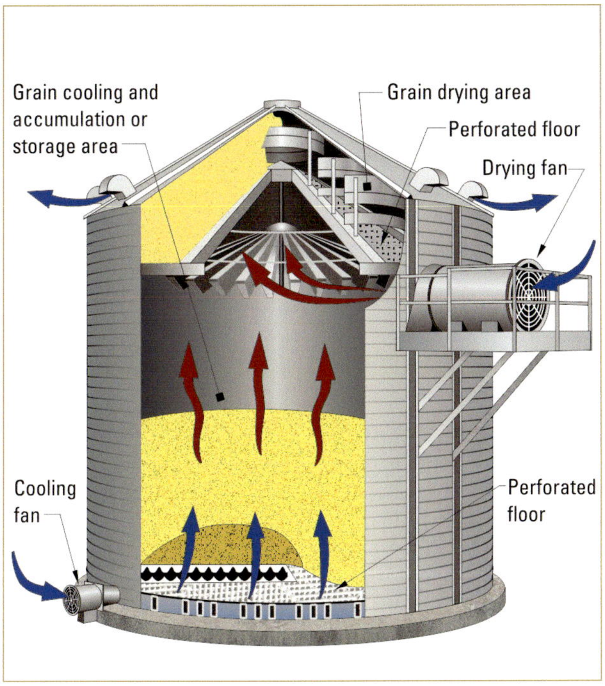


**Figure 4-13. Roof bin-batch dryer.**
A roof bin-batch dryer can dry grain for several storage bins, which improves effectiveness and reduces system cost.

Hot, dried grain drops onto a full perforated floor for cooling. Layers build up and the cooling front continues to move through the grain mass removing additional moisture through evaporative cooling. Hot grain is cooled with a separate fan while the drying chamber at the top is refilled with wet grain, and drying starts again.

The space between the grain being cooled and the drying floor is the hot air plenum. Exhaust air from the cooling grain batch is mixed with the heated drying air, decreasing drying efficiency somewhat. In older bins, fans with heaters are located at the top of the bin for moving hot air. In newer models, fans and heaters are located at the bottom for easy maintenance, with the heated air being ducted to the top.

There are few bottlenecks in handling dried grain with roof bin-batch dryers. An entire batch can be dropped from the drying area to the cooling area in less than 1 minute. However, loading requires more time. Use high-capacity loading conveyors to minimize delays. Automatically controlled low-capacity conveyors can be used to move dried, cooled grain to storage. Up to four batches can be dried per day.

### Continuous-flow bin dryer

Continuous-flow bin dryers are similar in design and set-up as on-floor bin-batch dryers except that the sweep auger is replaced with a single- or double-arm tapered screw auger that removes an 8 to 12 inch bottom layer of dried grain intermittently. These dryers are also known as continuous counterflow bin dryers because the downward grain flow and upward airflow are counter current to each other.

The hot, dried grain is transferred based on a timer and/or moisture sensor that will automatically turn on the tapered on-floor screw auger(s) and move the grain from the drying bin to bins equipped for in-bin cooling and storage. Transfer conveyors are sized to exceed the maximum expected drying capacity by 10 to 15%.

A major advantage of continuous-flow bin dryers is that wet-grain can be added to the top grain mass at any time which eliminates the need for a wet-grain holding bin. As a result, the grain depth in the bin dryer varies as a function of the drying rate and harvest rate.

Some continuous-flow bin dryers are equipped with stirring equipment which mix the upper portion of the grain mass and improve airflow from the drying layer. Stirring equipment is shortened so the drying layer of 4 to 6 feet remains undisturbed. For efficient operation it is important for continuous-flow bin dryers to be equipped with an effective spreader to keep the grain surface leveled as best as possible in order to ensure even airflow across the bin diameter.

Grain recirculation equipment is used only when the last batch of grain is dried and stored in the continuous-flow drying bin. If grain is stored in the drying bin, a perforated wall liner or evenly spaced perforated tubes are recommended to minimize condensation and subsequent caking of grain to the walls.

### Drying capacity

Drying capacity for all high-temperature bin dryers varies considerably with grain type, initial and final grain moisture content, loading and unloading time, airflow, drying air temperature, outdoor air conditions, and whether the grain is cooled in the bin. Changing any of these factors can either increase or decrease the drying capacity of the bin dryer.

The specific conveying equipment used can affect the amount of time required for loading and unloading the bin dryer. Decreasing the loading and unloading time reduces delays and increases available drying time and capacity.

Continuous-flow bin dryers require little delay time for conveying because grain is loaded and unloaded as drying continues. Bin-batch dryers require more time for conveying grain, which reduces available drying time and overall drying capacity. A roof bin-batch dryer can drop dried grain into the bottom of the bin for cooling in less than a minute, but requires similar time for unloading as other bin dryers. Minimize loading and unloading delays in roof and on-floor bin-batch dryers with high capacity conveying equipment.

As discussed in the fan section of this chapter, relative drying capacity (bu/hr) is directly proportional to airflow (cfm/bu). As airflow increases, drying capacity increases. Deeper grain depths increase airflow resistance, which reduces the amount of air delivered to the bin, decreasing drying capacity. Consider high-temperature bin dryers that are shorter than the storage bins, as illustrated in Figure 4-14. Replace lost storage capacity in the drying bin with taller storage bins.

Roof bin-batch dryers dry grain in batches 2.5 feet deep. On-floor bin-batch dryers without stirring often dry grain in batches up to 4 feet deep. Continuous flow and on-floor bin-batch dryers with stirring can be operated with grain depths of 9 feet or more, but have higher drying capacities at shallower depths. Drying depths up to 6 feet are preferred with continuous-flow and on-floor bin-batch dryers with stirring.

In addition to using shallower grain depths to increase drying capacity, adding another fan and heater can increase airflow and drying capacity, but doubling fans will not double drying capacity. For details, see Table 4-4. If more than one fan and heater are installed on a bin, use the same type, size, and brand to minimize uneven air distribution and drying. Table 4-7 shows the effect of airflow rate on drying capacity.

Increasing drying air temperature also increases drying capacity. Evaluate your situation and select the drying air temperature that fits your needs but do not exceed maximum recommended values. Consider grain type, available equipment, operating procedures, and previous experience with high temperature drying.

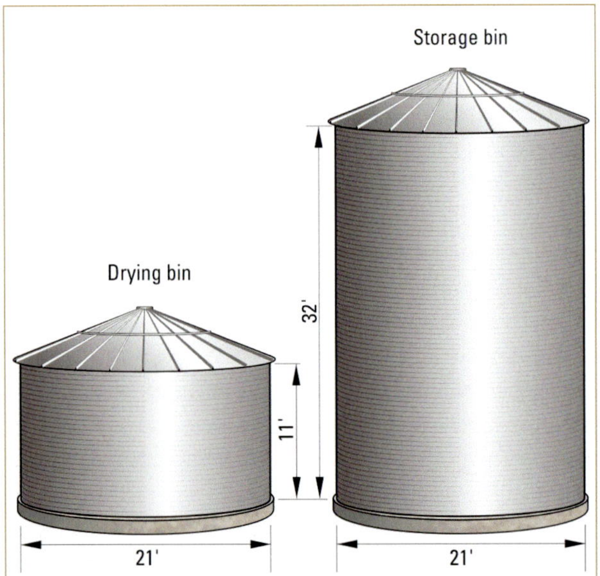


**FIGURE 4-14. Relative bin sizes for drying versus storage.**

**Table 4-7. Estimated capacities of bin dryers for maximum, practical airflow.**

These are estimated capacities to dry shelled corn from 25% to 15% including moisture (about 2%) removed in cooling.
Values assume maximum practical drying fan airflow, 180°F drying air temperature, and 4 feet grain depth.

| Bin dia. ft | Airflow cfm | Drying capacity | |
|---|---|---|---|
| | | bu/hr | bu/24 hr |
| 18 | 12,000 | 83 | 2,000 |
| 21 | 16,500 | 117 | 2,800 |
| 24 | 21,250 | 150 | 3,600 |
| 27 | 27,250 | 192 | 4,600 |
| 30 | 33,750 | 238 | 5,700 |
| 33 | 40,250 | 283 | 6,800 |
| 36 | 48,000 | 338 | 8,100 |
| 40 | 59,000 | 417 | 10,000 |

Consider using delayed cooling methods to increase drying capacity, and reduce drying costs and stress cracks. See the section on Grain Cooling later in this chapter.

Table 4-7 shows the maximum, potential drying capacity for high-temperature bin dryers with grain depths up to 4 feet. This table assumes minimum grain loading and unloading times, maximum practical fan capacity, 180°F drying air temperature, and no grain cooling in the drying bin. These values are the absolute maximum practical drying capacities under ideal conditions. It is impractical to expect these airflow rates and drying capacities at grain depths greater than 4 feet. You should not expect your drying system to achieve these capacities. They are presented as examples for comparison between different bin diameters.

Most high-temperature bin dryers are not selected or equipped for the maximum potential drying capacity. Many high temperature bin dryers will provide a drying capacity of about 1/2 the capacity shown in Table 4-7. When selecting the drying capacity for a bin dryer, consider cost, required capacity, operating requirements, and potential for expansion.

## Safety

Fires can occur in high-temperature bin dryers and are more common with some grains, especially sunflower and grain sorghum. Fans and burners require safety controls such as thermostats, high-temperature limit switches, airflow switches, and flame detectors. Control burner heat output levels from high-temperature bin dryers with modulating thermostats. Check safety and operating controls daily for proper operation. **Constantly** supervise sunflower and grain sorghum drying at all temperature settings. Fires can result from trash sucked into the fan and blown through the flame. Clean trash around and in the bin dryer daily. Foreign material and fibrous trash that hangs up on brackets and braces catch fire due to hotspots in dryers caused by an unevenly heated airflow downstream from the burners.

Cleaning grain that has a lot of trash will reduce fire risk. Consider building a vertical air intake around the fan and/or duct the fan intake away from the bin dryer. Size the duct for three times the fan's cross-sectional area to reduce air velocity in the air intake duct entrance.

## High-Temperature Self-Contained Dryers

High-temperature self-contained dryers can be manual batch, automatic batch, or continuous flow. All drying equipment such as gas plumbing, vaporizers, burners, blowers (fans), sensors, and control consoles are built into the dryer by the manufacturer. Filling and unloading equipment is typically not built into the dryer.

Some self-contained high-temperature dryers are movable and some are used as portable dryers; however, portable dryers, powered using a tractor P.T.O. are rarely used today. On-farm grain dryers commonly used in North America can be categorized as low-profile (horizontal - one level) dryers, stacked (multi-level) dryers and tower (vertical) dryers. However, there are other types of dryers such as mixed-flow (rack type) dryers that may not fit those categories. Stacked dryers can be single module/single fan units, which are stacked to make dual/multiple modules with dual/multiple fans. Tower dryers can be either modular (major component parts are in modules, which are assembled on site) or single tower systems assembled as one vertical unit.

These dryers are built to dry and cool grain in the dryer. However, they can be operated to eliminate cooling in the dryer (full heat) and move hot grain to a separate bin for cooling. Delayed cooling outside the dryer increases dryer capacity, reduces drying cost, and reduces stress cracks. Wet-grain holding and cooling method have to match drying capacity and grain flow. For more detail, see the section on grain cooling later in this chapter.

With on-farm dryers, fuel, electricity, and grain handling equipment are required at each drying site. Dryer burners can burn either liquid propane (LP) or natural gas (NG). LP is the most commonly used fuel on Midwest farms. Newer models of dryers have touchpad control screens with programmable logic controllers (PLC); moisture sensors; variable-frequency-drive (VFD) AC motors for efficient discharge of dry grain; energy-saving, heat reclaiming systems; and grain exchangers that enable more uniform heating of grain and a lower moisture differential between kernels in the drying column.

### Dryer selection

Consider filling, drying, and unloading requirements when selecting a high-temperature self-contained dryer. Dryer type affects the selection of wet-holding bins, conveying equipment, and grain cooling method. Drying speed, labor input, and required supervision during operation are specific to each dryer type. Consider your situation, including available labor during the next 5 to 10 years, and available conveying, wet-grain holding, and grain cooling equipment needed before selecting a high temperature self-contained dryer.

### Manual batch dryer

Manual batch dryers are typically selected for drying relatively small volumes of grain per year. With the increase in row crop acreage, very few of these dryers are used on U.S. farms today. An advantage of manual batch dryers is portability. They can be used in several locations, including in-field drying if they are PTO driven. Only the dryer, LP gas tank, and perhaps one conveyor need to be moved. Disadvantages are limited capacity, and the need for more labor and supervision for loading, drying, and unloading each batch.

### Automatic batch dryers

Automatic batch dryers operate like manual batch dryers, except loading and unloading are automatically controlled and require wet-holding and storage bins to be tied-in with the dryer. Automatic batch dryers have lower capacities than continuous-flow dryers, and are not very common on U.S. farms today.

Batch dryers do not dry grain during filling and unloading, so minimize downtime with high-capacity grain conveyors. A gravity-flow wet-holding bin is often installed over the dryer to provide fast loading. Use a high-capacity conveyor, such as a bucket elevator, for fast dryer unloading.

Consider delayed grain cooling in a cooling bin to eliminate the cooling cycle. This will increase dryer capacity and reduce drying costs and kernel stress cracks. For more information on this subject, see the section on grain cooling later in this chapter.

### Continuous-flow dryers

Continuous-flow dryers are high throughput systems. In the U.S. they are mostly cross-flow type dryers that consist of parallel perforated metal sheeting that form a column that is typically 10 to 14 inches wide to hold grain and allow it to flow by gravity from the top of the dryer to the bottom outlet. The drying air flows across the column with the hottest air in contact with the hottest grain next to the inside perforated sheet of the column. Columns are formed either into horizontal baskets or cylindrical towers.

More recently mixed-flow type dryers are making a comeback in the U.S. They consist of a taller enclosure equipped with racks that are spaced to allow grain to flow passed them by gravity from the top of the dryer to the bottom outlet. The drying air enters multiple racks from one side all at once and exits through the next set of racks above and below to the opposite side of the dryer. This creates a mixed-flow air pattern through the grain with some air flowing across, concurrent and counter to the grain. These dryers used to be popular in the U.S. before they were taken over by less expensive crossflow dryers. They remain popular in Europe, Asia, and Latin America, and a number of U.S manufacturers are offering new models across a range of capacities for on-farm and commercial applications. They create a more uniform drying effect and are thus preferred for process drying and specialty grain applicators such as for rice and food corn.

Stacked, cross-flow grain dryers have grain-column drying baskets with structural frames designed for modular units to be stacked one above another. They can operate in automatic-batch or continuous-flow modes. Stacked dryers can be single fan/single module units such as the example shown in Figure 4-15, or stacked to make dual/multiple or modules with dual/multiple fans and burners as shown in Figure 4-16.

In a single fan/single module dryer, (Figure 4-15), the fan is attached to a burner unit with both operating during the drying mode. In batch-mode, the burner is turned off once the grain reaches within about 0.5 percentage point of the desired final moisture content. The fan continues to blow ambient air through the grain column to cool the grain within 5 to 10°F or less of the ambient temperature which also removes the final amount of moisture.

In continuous-flow mode, the burner and fan stay on continuously and the grain leaves the dryer hot at about 1.5 to 2.5 percentage points above the desired final moisture content. The hot grain is transferred to a steeping and cooling bin for a process that is described in the section on grain cooling later in this chapter.

In a dual fan/single module dryer, either both fans or only the top fan are attached to burner units. When operating in heat-cool mode, the top fan and burner remain on to continuously dry the grain in the top two-thirds of the grain column to within about 0.5 percentage point of the desired final moisture content. The bottom fan remains on to continuously blow ambient air through the lower third of the grain column to cool the grain to within 5 to 10°F or less of the ambient temperature which also removes the final amount of moisture.

In full-heat mode, the top and bottom fans and burners both remain on continuously, and the grain leaves the dryer hot at about 1.5 to 2.5 percentage points above the desired final moisture content. The hot grain is transferred to a steeping and cooling bin.

**FIGURE 4-15. Single fan/module automatic-batch or continuous-flow dryer.**
*Photo courtesy of Sukup Engineering.*

In a multiple fans/multiple modules stacked dryer, typically all fans are attached to burner units. When operating in heat-cool mode, the bottom one or two fans serve as cooling fans. When operating in full-heat mode, hot grain is transferred to a steeping and cooling bin.

Wet grain is automatically fed from a wet-holding bin to the top of the dryer using a bucket elevator or inclined conveyor either continuously or in frequent intermittent cycles. The grain is distributed evenly across the length of the dryer by a fill-level conveyor. Grain flows downwards by gravity. Metering rolls at the bottom control the discharge of dry grain from the dryer based on a moisture feedback control system in the discharge spout. Size the loading and unloading conveyors 10 to 15% larger than the **maximum** grain drying capacity.

A disadvantage of column dryers is that the grain flowing along the inside perforated wall is exposed to the hottest drying air from the plenum. This can lead to overdrying of grain along that path and underdrying of grain flowing along the outer perforated wall exposed to lower temperature drying air with higher humidity. In the case of stacked dryer modules, grain mixing occurs as grain flows out of the bottom

a. Dual fans and single module dryer.

b. Four fans and three-stack modules dryer.

**FIGURE 4-16. Dual fans/single module and multiple fans/modules automatic-batch or continuous-flow dryers.**
*Photo courtesy of Sukup Engineering.*

of the top module into the top of the next module as illustrated in Figure 4-17. The effect of mixing drier grain from the inside of the column with wetter grain from the outside of the column mitigates overdying but does not eliminate it.

Some dryers use grain exchangers or grain turns, which are mounted in the grain column about half way down the drying section. These units split the grain column in half and move the inner grain against the outside of the column and the outer grain against the inside of the column (closer to hot plenum air) as illustrated in Figure 4-18. This will ensure more uniform distribution of both temperature and moisture in kernels across the drying column.

Another feature of some longer low-profile horizontal dryers is to place fan/burner units on both ends of the plenum. This results in producing a more uniform temperature across the length of the plenum. The merging of heated air streams from opposite ends of the heat plenum provides excellent mixing of heat throughout the plenum while equalizing static pressure in all corners of the plenum. This reduces low air volume and cooler temperatures near the front and high airflow with higher temperatures at the back of the plenum.

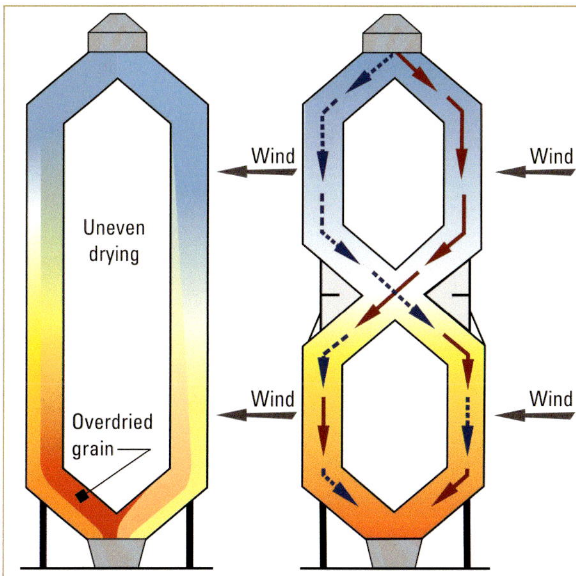


a. b.

**FIGURE 4-17. Illustration showing how grain mixes in the section between two stacked dryer modules, (b) leading to more even drying compared to a taller drying column (a).**

Tower dryers consist of a circular column structure that can range in height from 40 to 130 feet. Grain is fed from the top of the dryer and flows down by gravity in concentric cylindrical columns, shown in Figure 4-19. Heated air is blown by high efficiency burners/fans from the bottom up, into the plenum and across the grain column. The drying section occupies about two thirds of the tower; the cooling section the bottom one third.

Cooling is achieved by suction of ambient air by the dryer fan(s), which are typically installed inside the bottom portion of the tower, through the column of hot grain. As a result of cooling the grain, the air is warmed up before it is pushed by the fan through the burner and up into the drying plenum. The resulting heat recovery from the hot grain and pre-heating of the drying air before the burner results in substantial energy savings for tower dryers and improves energy efficiency by about one third.

Most tower dryers today have grain exchangers that function as illustrated in Figure 4-18. This enables more uniformity in kernel moisture than conventional tower dryers.

Ground-level wet-holding bins with a bucket elevator to keep the dryer filled and a second bucket elevator to move dried grain to cooling or storage bins are common.

Figure 4-20 illustrates the modular tower design of a mixed-flow dryer and its internal structure of honeycombed, triangular baffles installed in a staggered configuration. Like other dryers, they can be operated in full-heat or heat/cool modes.

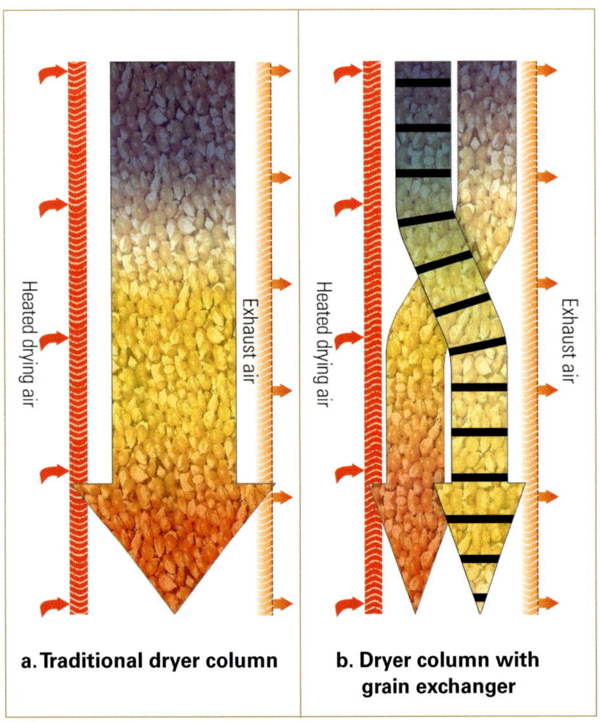


**FIGURE 4-18. Illustration showing how grain exchangers in column dryers work.**

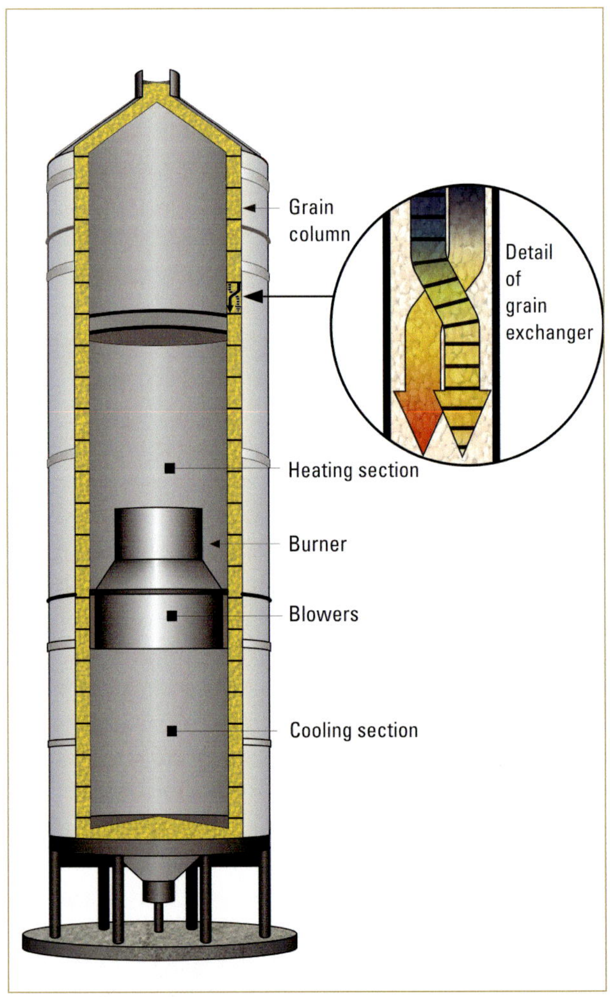


**FIGURE 4-19. Illustration of a crossflow column tower dryer.**
Grain exchanger moves grain from inside of the column to the outside.

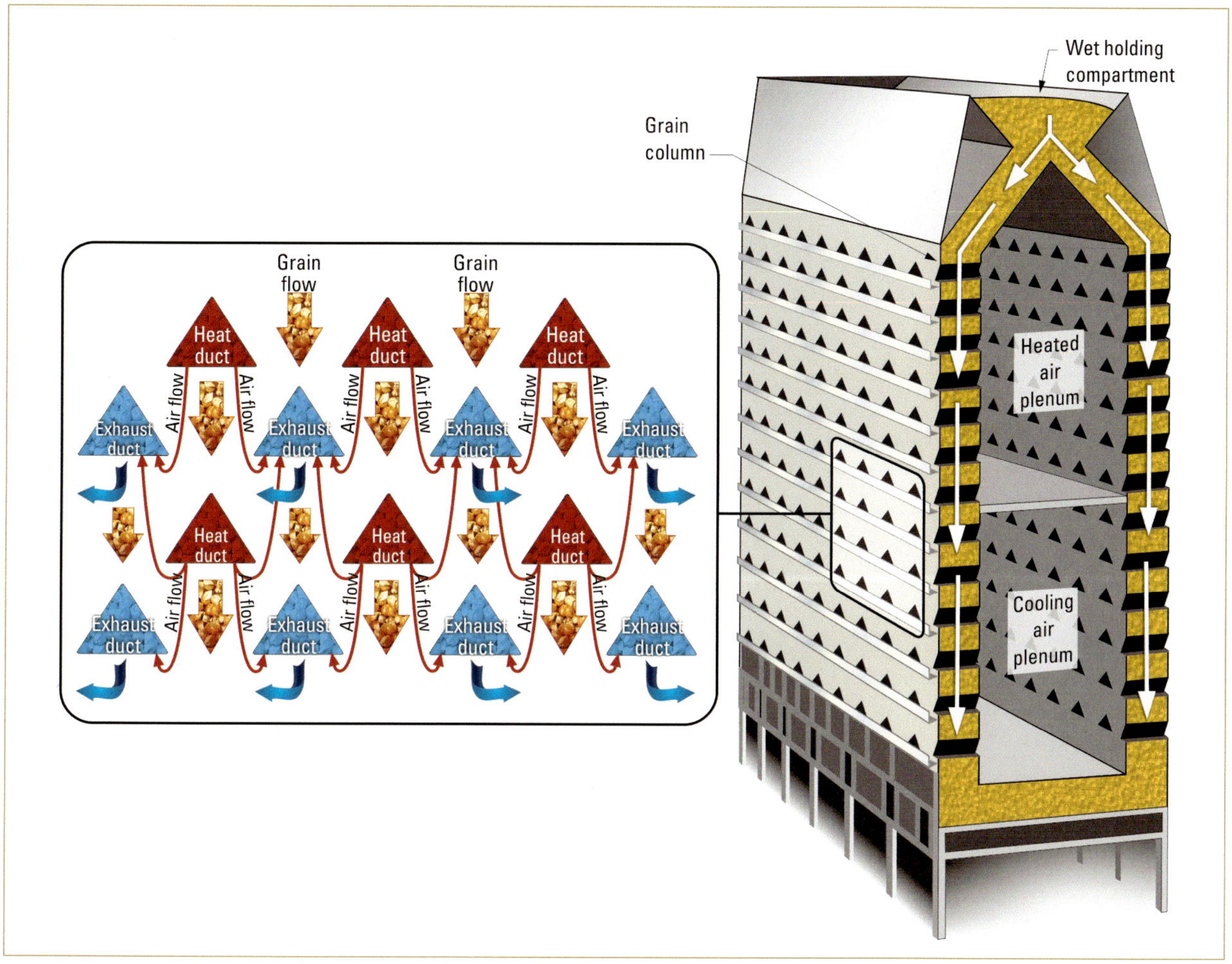


**FIGURE 4-20. Illustration of a mixed flow tower dryer.**

### Drying capacity

The capacities of high-temperature, self-contained dryers are often quoted for ideal drying conditions—such as warm, low humidity weather, and hybrids that release their kernel moisture easily.

However, ideal drying conditions rarely occur in the real world. The realistic capacity is often 70% to 80% of the advertised value. In general, capacities of grain dryers are reported in wet bushels per hour (bu/h) and are typically rated based on a 5 percentage points (20 to 15%) and 10 percentage points (25 to 15%) moisture removal. The drying capacity for 10 percentage point moisture removal is about 60% of the 5 percentage point moisture removal. The estimated capacity is generally based on an outdoor temperature of 50°F and air relative humidity of 50 percent. Cooler temperatures and higher humidities will reduce the drying capacity.

The rated drying capacities (bu/h) of single fan/module dryers, like the one shown in Figure 4-15, range from 200 to 700 bu/h for 10 percentage points moisture removal and 300 to 1,100 for 5 percentage points moisture removal. Drying capacities (bu/h) of multiple fans/ two modules continuous-flow dryers range from 800 to 1,600 for 10 percentage points moisture removal and 1,500 to 2,500 for 5 percentage points moisture removal in full heat mode. For heat/cool mode, the drying capacities (bu/h) are less: 900 to 1,700 for 5 percentage points moisture removal and 600 to 1,100 for 10 percentage points moisture removal. Drying capacities (bu/h) for multiple fans/multiple modules stacked continuous-flow dryers for full heat mode range from 2,300 to 3,700 for 5 percentage points removal and 1,400 to 2,400 for 10 percentage points moisture removal. For heat/cool mode, the drying capacities range from 1,400 to 2,400 bu/h and 900 to 1,400 bu/h for 5 and 10 percentage points moisture removal, respectively.

For tower dryers, drying capacities range from 800 to 7,000 bu/h for on-farm systems, and up to 10,000 bu/h for commercial elevator dryers.

A well-designed system can dry 24 hours a day, but many farmers operate their combines and dryers 10 to 16 hours a day, so they can supervise drying, particularly with widely varying grain moisture content. High-temperature drying for sunflowers, grain sorghum, and soybeans requires constant supervision because of potential fire hazards. However, some drying efficiency is lost every time a continuous-flow dryer is allowed to run empty or started back up.

To estimate required drying capacity, divide the maximum expected bushels harvested per day by hours of dryer operation per day.

**For example:** 300,000 bu harvested in 20 days, requires a dryer capacity of 15,000 bu/day (300,000 bu/20 days). To operate the dryer only 10 hours per day, dryer capacity must be 1,500 bu/hr (15,000 bu/day/10 hr/day). To operate 15 hours per day, capacity can be 1,000 bu/hr. If the dryer runs automatically 24 hours per day, 625 bu/hr capacity is adequate.

Early season drying will be slower than later season due to higher grain moisture. However, later season weather may be more of an issue, which may keep corn moisture higher than expected due to slow field drying in bad weather, so adjust your drying capacity and hours of drying per day based on your experience with local harvesting conditions.

Plan the drying system so future annual drying capacity can be increased at least 50% (multiply by 1.5). For more flexibility, plan for doubling the drying capacity. This provides the opportunity to evaluate what might change within the probable life of the drying system. Drying capacity can be expanded by drying more hours per day, or by incorporating dryeration, in-bin cooling, or combination drying. A 2,000 bu/hr dryer with an in-dryer cooler operating 12 hours per day dries 24,000 bu/day. Drying 20 hours per day increases capacity to 40,000 bu/day. Converting the dryer to full heat operation with dryeration of hot corn transferred at 16.5 to 18% moisture content increases capacity by at least 60% to 38,400 bu/day with 12-hour drying or 64,000 bu/day with 20-hour drying. If the system includes low-temperature drying bins (combination drying) where additional drying, tempering and cooling of hot corn transferred at 18 to 20% moisture content is done outside the high-temperature dryer, then the overall drying system capacity can be increased to 60,000 bushel in 12 hours or 100,000 bushel in 20 hours.

### Energy conservation

Grain drying consumes a lot of energy, so save what you can.

**For example:** Dryeration, one method of delayed cooling, can cut 25 to 30% off energy consumption of high temperature drying while increasing drying capacity by 50 to 70% for 10 percentage points of moisture removed. In-bin cooling and combination drying with high-temperature dryers are other methods that can save similar amounts of fuel energy. For more details, see the grain cooling section later in this handbook.

Recovering cooling air and lower-heat zone, discharge air from column-type continuous-flow dryers can save 20 to 30% of drying energy. This much savings usually occurs when drying relatively wet grain during cold weather. Some dryers recover and recirculate all discharge-cooling air, and others recover all cooling air and some of the discharge air from the lower portions of the heated air section. Good air recovery systems are practical energy savers with very little loss in dryer capacity. Several manufacturers have energy conservation kits available for older model dryers. Most tower dryers utilize heat recovery through vacuum cooling with fans that pull ambient air through hot, dry grain prior to introducing it through the burners. Heat recovered from hot grain is recycled back for drying wet grain, thus saving energy.

Recovering discharge air from the lower part of a full-heat (no in-dryer cooling) continuous-flow dryer adds up to 20% more energy savings. Actual energy savings by air recovery depends on dryer airflow (cfm), final grain moisture content, moisture content reduction in the dryer, outdoor and drying air temperatures, and the proportion of total dryer air-flow that is recovered. Exhaust air from the bottom 25 to 30% of the heat zone is usually warm, low-relative-humidity air. It has additional potential to reduce grain moisture content when reheated and passed through higher moisture grain in the drying section.

## GRAIN COOLING

Grain dried in any heated-air dryer must be cooled. Grain can be rapidly cooled immediately after it is dried, or delayed cooling methods can be used to reduce fuel costs, increase dryer capacity, and reduce stress cracks. The cooling method can affect the type, operation, and management of the dryer system. Consider the cooling method when selecting a dryer.

With delayed cooling, hot grain at 120 to 140°F is usually transferred from a high-temperature dryer to a separate bin to be steeped (or tempered) and cooled. Additional handling equipment and cooling bins with a fan and perforated floor are needed for delayed cooling. In-bin cooling, dryeration, and combination high/low temperature drying are the three delayed cooling methods. Delayed cooling effectiveness increases as cooling air temperature decreases. The minimum drying air temperature to consider for delayed cooling is 120°F, but hot grain from a 220°F dryer temperature will have much more heat energy for evaporative cooling. Hot grain will also have much more heat stress relieved by tempering for 1 to 4 hours or more before cooling. Tempering allows the internal kernel heat and its vapor pressure to sweat, that is move the moisture remaining at the core of the kernel to the outer kernel layers. This relaxes the stress build up inside the kernels due to uneven moisture distribution inside the kernels. In addition to moisture redistribution and stress relaxation inside each kernel, tempering also causes moisture to equalize between kernels. This is especially important in column dryers where kernels flowing along the inside plenum screen are overdried versus kernels along the outer screen. Moisture is removed more easily and efficiently with dryeration cooling than during high-speed cooling in the lower part of a conventional heat and cool continuous-flow dryer.

Condensation of moisture inside the cooling bin on the wall and roof typically occurs with delayed cooling using positive pressure airflow. Fines and bees wings will stick on bin walls with condensation, initiating crusting and making discharging the bin a challenge. When ambient temperatures are below freezing, wet corn will freeze to the bin wall. Ensure that bins used for delayed cooling are equipped with an adequate amount of vent openings and roof exhaust fans to exhaust saturated, humid air. See MWPS-29, *Dry Grain Aeration Systems Design Handbook* for a guide on sizing vents and roof exhausters for intake and exhaust area.

## In-Bin Cooling

In-bin (or in-storage) cooling is the simplest delayed cooling method and is suitable with any type of high-temperature dryer. Figure 4-21 illustrates how grain can be dried in either a high-temperature self-contained dryer or a bin dryer and cooled in a separate storage bin. Compared with rapid in-dryer cooling, in-bin cooling can reduce fuel costs at least 10% and increase dryer capacity about 30%. For a comparison of relative operating costs, see Table 4-8. Attention should be paid to condensation particularly at ambient temperatures below 50°F.

The general operating procedure is to stop high temperature drying about 1 to 1.5 percentage points above the desired final moisture content. Transfer grain to storage and operate the fan(s) continuously to remove the extra moisture by evaporative cooling. Because cooling is not delayed long, cooling removes about 0.1 to 0.15 percentage points for each 10°F reduction in grain temperature (about half as much as dryeration where 2 to 2.5 percentage points are typically removed). After cooling, grain is usually stored in the same bin.

Start in-bin cooling as soon as half or more of the grain is in the cooling bin. In-bin cooling can be delayed if outside temperatures are above 60°F. If cooling is delayed, especially in cold weather, moisture condenses on the sidewalls and accumulates in the grain, which can cause spoilage.

Size the fan so cooling is at least as fast as drying. For example, for a 500 bu/hr drying capacity, size fans to cool at least 500 bu/hr when the bin is full. Determine the **minimum** airflow (cfm) by multiplying the desired cooling rate by 12 cfm/bu/hr. For a 500 bu/hr dryer, the desired cooling rate is 500 bu/hr and the minimum airflow is 6,000 cfm (12 cfm/bu/hr × 500 bu/hr).

The cooling fan must deliver at least 6,000 cfm at the maximum grain depth. However, the airflow rate should not be more than 0.5 cfm/bu. Too high of an airflow cools the hot grain too fast and decreases the amount of moisture removal due to evaporative cooling. You will need to determine the static pressure to select a fan. State Extension agricultural engineers and knowledgeable grain drying equipment dealers can help estimate static pressure for specific in-bin cooling application, or use the University of Minnesota Fan Selection for Grain Bins web app that can be accessed at http://webapps.bbe.umn.edu/fans/.

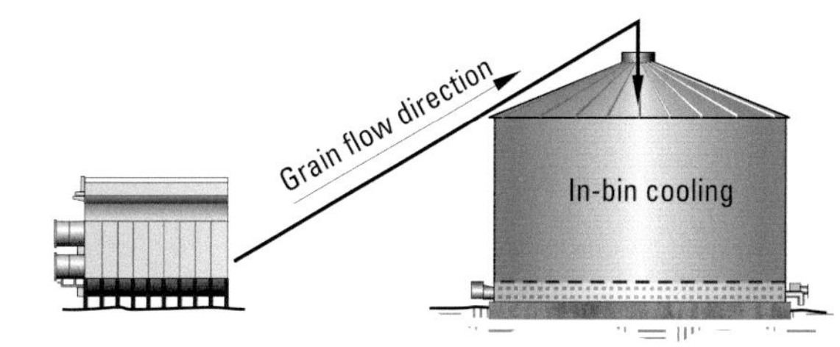


**a. With high temperature self-contained dryer**

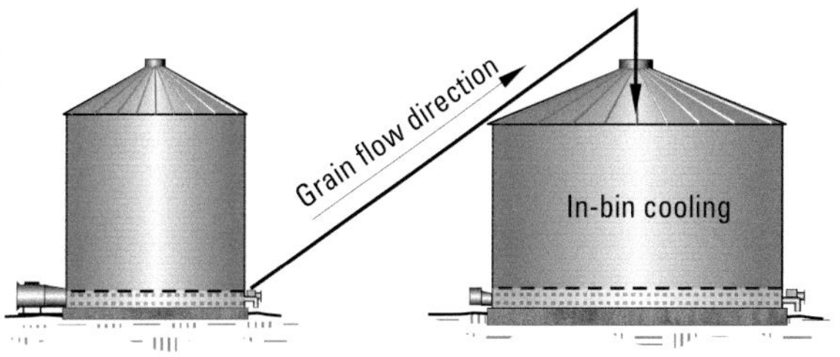


**b. With high temperature bin dryer**

**FIGURE 4-21. In-bin cooling.**
Grain can be dried in a high-temperature column or bin dryer and cooled in a separate storage bin.

An aerated storage bin can be used for in-bin cooling if filled gradually. With 0.1 cfm/bu aeration capacity when full, fill the bin and cool the hot grain over 6 or more days. With 0.2 cfm/bu aeration capacity, fill the bin and cool the got grain in 3 or more days. Ideally, a cooling bin should be equipped with a grain spreader to fill the bin with hot grain as evenly as possible. Level fill assures uniform airflow and cooling effect.

Bins with fully perforated floors provide the best air distribution for in-bin cooling. Partly perforated floors or ducts are adaptable, but should be avoided because air movement through the grain will be less uniform. With partial floors or ducts, it is safer to partially cool the first 2 to 3 feet of grain before adding more hot grain.

## Dryeration—Batch Tempering Cooling Systems

Dryeration is the most energy efficient, quality maintaining drying method. It involves both tempering and delayed cooling. Compared with immediate, rapid cooling in the dryer, dryeration reduces fuel costs 15 to 30%, increases drying capacity 50 to 70%, and results in fewer stress cracks, less brittleness, improved color, and better millability of grains. See Table 4-8 for a comparison of drying capacity and energy costs.

The two dryeration systems used in the U.S. are batch and continuous-flow dryeration. Batch dryeration typically utilizes two hopper bins used on alternate days to accomplish the tempering and cooling processes while continuous-flow dryeration systems utilize one bin to accomplish both tempering and cooling by continuously adding hot grain to the top and unloading cool, dry grain from the bottom.

With batch-dryeration, hot grain at 2 to 2.5 percentage points above the desired final moisture content is moved to the tempering bin immediately after drying. Cooling is delayed at least 4 hours and preferably up to 12 hours for tempering. This allows time for the moisture content to equalize in the kernels and among kernels which reduces physical stress during cooling and final moisture removal. After cooling, grain in the dryeration bin is transferred into storage. Proper tempering and cooling in a dryeration bin can remove 2 to 2.5 percentage points of moisture content. Figure 4-22 illustrates the typical operating procedure for a two-dryeration batch bin system.

An important dryeration concept is that all hot grain delivered to a dryeration bin must temper at least 4 hours before it is cooled.

*Two possible filling and cooling methods assure that all grain tempers the required time:*

- Fill the dryeration bin with one day's dryer operation (up to 24 hours) and delay starting the cooling fan until after the last hot grain is transferred. The bottom grain is tempered for 8-12 hours before starting the cooling fan. If the bin cools in 8 to 12 hours, the top half of the grain tempers 8 to 12 hours before the cooling front reaches it. Once cooled, transfer the grain to storage.

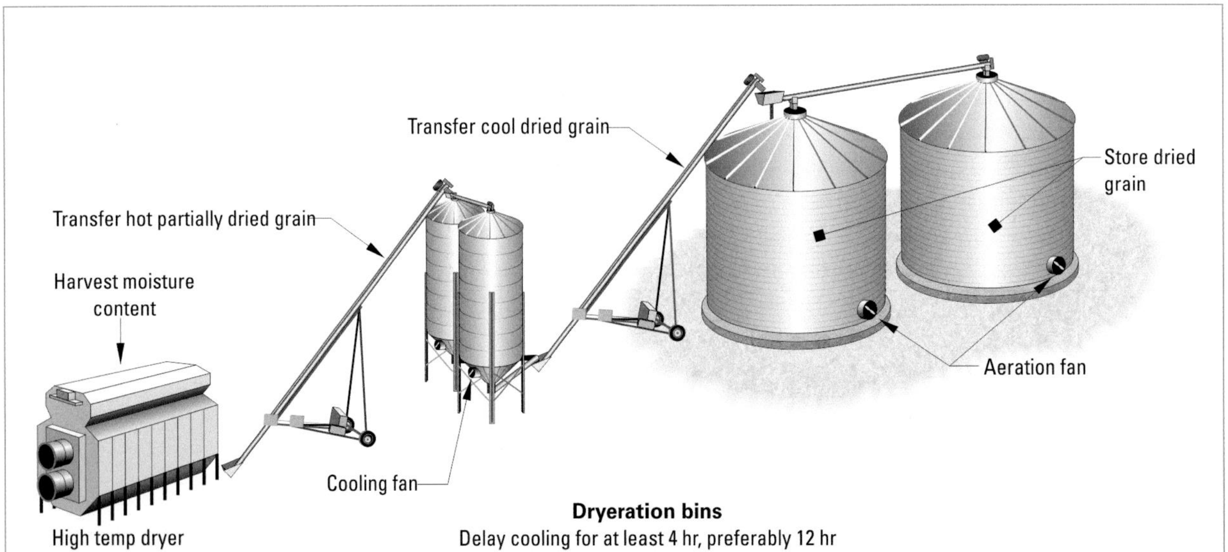


**FIGURE 4-22. Example of the typical operating procedure for a two-dryeration batch bin system.** The high temperature column dryer transfers hot grain into one dryeration bin while grain in the second dryeration bin is allowed to temper, is then cooled and then emptied into a storage bin. The cycle repeats.

- Begin filling the dryeration bin, and delay starting the cooling fan at least 4 to 6 hours after the first hot grain enters the bin. Tempering and cooling continues until the bin is filled and until the cooling front has moved through the hot grain entirely. Once cooled, transfer the grain to storage.

With dryeration, air is always moved up through the bin to cool grain. As the grain cools, the air warms to the grain temperature. As the air warms, it picks up moisture from the grain. This warming and drying happens over a short distance in the grain, forming a cooling front that gradually moves through the bin. Air moving through warm grain ahead of the cooling front does not significantly change the warm grain temperature or moisture content, so warm grain still tempers even though air is moving through it. If air moves upward, the cooling fan can run while hot grain is being added. However, the last grain into the bin should temper at least 4 hours before being cooled. Note that using negative airflow is considered undesirable because hot grain added to the bin will heat the air and limit cooling.

The cooling and drying effect of dryeration reduces grain moisture content about 0.2 to 0.25 percentage points for each 10°F reduction in grain temperature. So, the high-temperature dryer can be stopped when grain moisture content is 2 to 2.5 percentage points higher than the desired final moisture content.

The fill level and the cooling front both advance upward in the bin. If the cooling front overtakes the filling rate, the last grain loaded into the bin will not sufficiently temper before cooling. The rate at which the cooling front moves depends on the airflow rate. If the cooling front moves faster than the filling level, reduce the airflow rate with intermittent fan operation. A percentage timer, time clock, or other electrical control device can operate the fan intermittently.

For dryeration, provide airflow of at least 0.5 cfm/bu but no more than 0.75 cfm/bu when the bin is full. The higher the airflow, the faster grain cools, but the faster cooling produces lower drying efficiency: 0.5 cfm/bu tempers and cools hot grain in about 24 hours—0.75 cfm/bu in about 16 hours.

**For example:** For a 6,000-bushel cooling bin, the minimum required airflow is 3,000 cfm (6,000 × 0.5 cfm/bu).

You can also base required airflow on dryer capacity. Tempered hot grain is usually cooled at the same rate that the dryer fills the bin, so cooling rate and airflow (cfm) depend on dryer capacity. A good rule of thumb is to multiply the maximum dryer capacity (bu/hr) by 12 cfm/bu/hr to get minimum airflow (cfm).

**For example:** If the maximum dryer capacity is 1,000 bu/hr, the required minimum airflow is 12,000 cfm (1,000 bu/hr × 12 cfm/bu/hr).

When grain is being tempered and cooled with dryeration, condensation usually accumulates on and in grain near the bin walls, particularly during cold weather. Tempering for more than 16 hours in below freezing weather can lead to frozen condensation. Condensation will cause problems if grain is stored in the dryeration bin. But, when grain is transferred from dryeration to storage, the process of unloading and conveying mixes small quantities of wet grain with dry grain, so it stores well. Dryeration requires a convenient, well planned grain transfer system. Get design help from your state Extension agricultural engineer, a consulting engineer, or a commercial vendor.

Hopper-bottom bins with fans and ducts, and flat-bottom bins with fans and perforated floors are both used for dryeration. With flat-bottom bins, a cone of dry grain can be left in the bin between batches to form a hopper shape. Check the cone regularly for surface crusting that might block airflow. If grain will be stored in the bin, remove the cone with a sweep auger before the season's last filling to remove any accumulated moisture, fines or moldy grain.

Size cooling bins to store at least the maximum dryer output for the cooling cycle—usually 16 to 24 hours. Remember that dryer capacity increases 50 to 70% with dryeration. Consider oversizing cooling bins up to 50% (1.5 times daily drying capacity) to provide flexibility if drying capacity increases.

**For example:** A high temperature dryer dries up to 16,000 bushel in a 16 hr/day drying period. Provide two cooling bins with a capacity of at least 24,000 bushel each (16,000 bu × 1.5). The extra capacity provides flexibility to change the drying operation. If drying is extended to 24 hours per day, then 24,000 bushel of capacity is needed.

## Alternative Dryeration Hopper-Bin Management Systems

Dryeration works with batch tempering/cooling using two dryeration bins and additional storage bins. The original design was to use an alternating two-day cooling bin cycle for loading, tempering, cooling, and unloading. Example 4-2 uses two bins which each hold one day's harvest and drying. Example 4-3 uses only one large bin, drying a day's harvest in about 2/3 of a 24-hour day, with larger fans and higher unload conveyor capacities. Tempering and cooling cycles are shorter, using larger fans, and higher conveying rates, where dryeration is completed once every 24 hours.

Example 4-4 is for continuous 24-hour drying, using two bins with higher unloading rates, where each bin receives half of a day's harvest, but finishes its dryeration cycle in a 24-hour period. This allows use of smaller, lower cost bins, with larger fans and unload capacities.

The conventional dryer capacity in Example 4-4 is only 60% (300 bu/hr vs 500 bu/hr) of that in Examples 4-2 and 4-3 to demonstrate the power of dryeration to produce similar capacity with smaller equipment running full time.

**EXAMPLE 4-2. Dryeration—Dual Batch-bins**

**15-hr Drying**

**SYSTEM SET-UP**

Dryer, Heat/Cool = 500 bu/hr
All-Heat Dryer + 70% w/Dryeration
500 × 1.7 = 850 bu/hr

**SYSTEM SIZING**

*Drying:*
15 hr/day × 850 bu/hr = 12,750 bu/day
12 hr/day × 850 bu/hr = 10,200 bu/day

*Combine:*
12 hr/day; 7AM to 7 PM = 12,750 bu/day

*Wet Holding:*
12,750 – 10,200 = 2,550 bushels wet holding in bins, wagons, or trucks at end of the day

*Dryeration:*
Two 14,000 bu Hopper Bins (10% greater than calculated need)

*Aeration:*
Use FANS program to select fan to deliver 0.5 cfm/bu: 12,750 bu × 0.5 cfm/bu = 6,375 cfm

*Unload Conveyor:*
Dryeration unload at 1500 bu/hr = 12,750 bu / 1500 bu/hr = 8.5 hours

**PROCESS STEPS**

**Day 1—Bin #1**

1. Start loading Dryeration Bin #1 at 8:00 AM
2. Finish loading after 15 hours at 11:00 PM

**Day 2—Bin #1**

3. Start fan at 7:00 AM (top grain tempers 8 hours before cooling starts).
4. Stop fan at 7:00 PM after 12 hours of evaporative cooling.
5. Unload Bin #1 at 1,500 bu/hr, starting at 7:00 PM for 8.5 hours.
6. End unloading at 3:30 AM, Day 3 and Bin #1 is ready to receive hot corn first thing in the morning.

**Day 2—Bin #2**

7. Start loading Dryeration Bin #2 at 8:00 AM
8. Finish loading after 15 hr at 11:00 PM on Day 2.
9. Repeat same sequence as Day 1 for Bin #1, above. Bin #1 is loaded on the odd days and Bin #2 is loaded on the even days.

**EXAMPLE 4-3. Dryeration—Single Batch-bin**

**15-hr Drying at Higher Temperature**

**SYSTEM SET-UP**

Dryer, Heat/Cool = 500 bu/hr
All-Heat Dryer + 80% w/Dryeration
500 × 1.8 = 900 bu/hr

**SYSTEM SIZING**

*Drying:*
14.2 hr/day × 900 bu/hr = 12,780 bu/day
12 hr/day × 900 = 10,800 bu/day

*Combine:*
12 hr/day; 7AM to 7 PM = 12,780 bu/day

*Wet Holding:*
12,780 – 10,800 = 1,980 bushels wet holding in bins, wagons, or trucks at end of the day

*Dryeration:*

One 14,000 bu Hopper Bin
(10% greater than calculated need)

*Aeration:*

Use FANS program to select fan
to deliver 0.5 cfm/bu:
12,780 bu × 0.5 cfm/bu = 6,390 cfm

*Unload Conveyor:*

Dryeration unload at 3,000 bu/hr =
12,780 bu / 3,000 bu/hr = 4.3 hours

**PROCESS STEPS**

**Day 1**

1. Start loading Dryeration bin at 8:00 AM.
2. Finish loading after 14.2 hours at 10:15 PM.

**Day 2**

3. Start fan at at 10:15 PM (top grain tempers long enough before cooling front reaches the top).
4. Stop fan at 10:15 AM after 12 hours of evaporative cooling.
5. Unload bin at 3,000 bu/hr, starting at 10:15 AM for 4.5 hours.
6. End unloading at 2:45 PM.
7. Start reloading Dryeration bin at 3:00 PM.

**Day 3**

8. Finish loading at 5:15 AM.
9. Repeat same sequence as Day 1, above.

**EXAMPLE 4-4. Dryeration—Dual Batch-bins**

**24-hr Drying**

**SYSTEM SET-UP**

Dryer, Heat/Cool = 300 bu/hr
All-Heat Dryer + 70% w/Dryeration
300 × 1.7 = 510 bu/hr

**SYSTEM SIZING**

*Drying:*

24 hr/day 510 × bu/hr = 12,240 bu/day
12 hr/day × 510 bu/hr = 6,120 bu/day

*Combine:*

12 hr/day; 7AM to 7 PM = 12,240 bu/day

*Wet Holding:*

12,240 – 6,120 = 6,120 bushels wet holding in bins, wagons, or trucks at end of the day

*Dryeration:*

Two 7,000 bu Hopper Bins
(10% greater than calculated need)

*Aeration:*

Use FANS program to select fan
to deliver 0.5 cfm/bu: 7,000 bu × 0.5 cfm/bu =
3,500 cfm

*Unload Conveyor:*

Dryeration unload at 1,250 bu/hr =
6,120 bu/1,250 bu/hr = 4.9 hours

**PROCESS STEPS**

**Day 1—Bin #1**

1. Start loading Dryeration Bin #1 at 8:00 AM.
2. Finish loading after 24 hours at 8:00 PM.

**Day 1—Bin #2**

3. Switch to Dryeration Bin #2 at 8:00 PM.

**Day 2—Bin #1**

4. Start fan at 2:00 AM (top grain tempers 6 hours; before cooling starts).
5. Stop fan at 2:00 PM after 12 hours of evaporative cooling.
6. Unload bin at 1,250 bu/hr, starting at 2:00 PM for 5 hours.
7. End unloading at 7:00 PM. Bin #1 is ready to receive hot grain again.
8. Repeat same sequence as Day 1 for Bin #2 above.

**Day 2—Bin #2**

9. Finish loading at 8:00 AM (12 hours)
10. Start fan at 2:00 PM (top grain tempers for 6 hours before cooling starts).

**Day 3—Bin #2**

11. Stop fan at 2:00 AM after 12 hours of evaporative cooling.
12. Unload bin at 1,250 bu/hr, starting at 2:00 AM for 5 hours.
13. End unloading at 7:00 AM. Bin #2 is ready to receive hot grain again. Repeat same sequence as Day 1 for Bin #1, above. Bin #1 is loaded on odd days, and Bin #2 is loaded on even days.

## Dryeration—Continuous-Flow Tempering/ Cooling Systems

Continuous-flow dryeration systems utilize one bin (or holding structure) to accomplish both tempering and cooling by continuously adding hot grain to the top of the grain mass and unloading cool, dry grain from the bottom. Available systems consist of flat or hopper bottom bins, or a self-contained, low-profile horizontal, modular-stacked or vertical tower design.

A flat-bottom bin with a full perforated floor can be converted by adding a single- or double-arm tapered screw auger that removes an 8 to 12 inch bottom layer of cool, dry grain intermittently and an aeration fan that is sized to deliver no more than 0.5 cfm/bu at the operating grain depth. This is essentially the same set-up as a continuous-flow bin dryer except for the smaller-fan (and lower design airflow rate) and a greater grain depth. The upper half serves as the tempering zone and the lower half as a cooling zone.

In the cooling zone ambient air flows counter current upward through the hot corn exchanging heat and moisture through evaporative cooling. Once the "cooling" air reaches the hot grain temperature and 100% saturation, no further heat and moisture exchange occurs. From that level in the grain mass the air flows through the tempering zone where kernels "sweat" and relieve internal stress, equilibrate internal kernel moisture gradients, and equilibrate moisture among kernels.

The grain depth is controlled by lower and upper limit switches that maintain a minimum and maximum depth of the steeping and cooling zones. Below the limit the tapered unloading screw(s) stop removing grain from the bottom while the fan is allowed to continue evaporative cooling of the grain mass. Above the limit the transfer of hot grain from the dryer operating in full-heat mode would be stopped to avoid overflowing the tempering and cooling bin. The tapered screw auger is controlled automatically based on a temperature and moisture sensor installed in the underfloor unload conveyor.

The bin needs to be sized in diameter based on the lengths of commercially available tapered screw unload conveyors, and grain depths that allows for a minimum tempering time of 4 hours in the tempering zone of the grain mass while keeping up with the maximum capacity of the dryer operating in full-heat mode. Fan(s) are sized to deliver no more than 0.5 cfm/bu airflow rate to avoid the non-saturated, hot "cooling" air from encroaching into the tempering zone. An effective grain spreader to level-fill the continuous-flow dryeration bin is critical for uniform airflow through the cooling and tempering zones. Managing humid exhaust air exiting the bin to minimize condensation on the underside of the roof as well as in transfer conveyor is critical. Pneumetric conveying of hot grain from the full heat dryer to the continuous-flow dryeration bin is preferable.

In terms of hopper-bottom bins, the angle of the hopper needs to be steep enough (at least 45°, preferably 60°) so that the grain flows through and out of the bin as close to plug flow as possible. This avoids the need for a flat bottom bin and a tapered screw conveyor to mechanically unload the continuous-flow dryeration bin. In all other aspects, the same design, equipment sizing, control and operational process as for the flat-bottom bin applies to the use of hopper-bottom bins for continuous tempering and evaporation cooling of hot grain.

Hot grain can also be transferred into self-contained low profile horizontal, modular stacked or vertical tower units such as old dryers where hot grain can be tempered in the upper sections previously used for drying, and cooled in the bottom section but with smaller fans to a maximize the evaporative cooling effect. Perforated outside screens of column dryers in the upper suction should be covered with metal sheeting to avoid wind or fan blowing through columns and cooling hot grain in the tempering zone. Fans in tower dryers that are set up for section cooling would have to be down-sized and reversed to push cooling air out through the column.

Old mixed-flow rack-type dryers work well for continuous-flow dryeration as they are already fully enclosed. Racks in the drying section can be easily covered to assure cooling air is directed through the lower cooling section only. Some manufacturers have made specially designed towers commercially available that eliminated the racks in the tempering section but have them in the cooling section. Counter-flow coolers as used for pellet cooling in the feed industry can also be adapted easily for the continuous-flow dryeration of hot grain.

Fan and loading/unloading equipment sizing, and process control and operation is similar for the bin as well as self-contained dryeration system. Do not let cost or complexity discourage you from considering dryeration with high-temperature drying. Even though more equipment and management are required, the extra energy savings, greater dryer capacity, and improved grain quality more than offset the disadvantages. If you have **any storage, handling, or marketing problems with fines and broken grain**, seriously consider dryeration. Remember, dryeration increases drying capacity, decreases drying costs, and maintains grain quality. Corn dried and cooled in conventional, high-temperature dryers without energy saving or kernel mixing features will suffer substantial damage **each time it is handled!**

### Combination High-Temperature/Low-Temperature Drying

Combination high-temperature/low-temperature drying is discussed here as a cooling method because grain cooling is separate from the high-temperature dryer.

Combination drying is most adaptable for crops harvested too wet for safe, low-temperature bin drying. It is usually not used for wheat or soybeans but works well for corn. Above 22% moisture, grain goes first to the high-temperature dryer. Below 22%, it goes directly to the low-temperature bin dryer. Wet corn (above 22% moisture content) can be dried rapidly with high-temperature heated air to a moisture content of 20% or less. As the partially dried corn is delivered hot to a low-temperature drying bin, start the drying fans immediately. Bin fans provide much lower-velocity cooling air than the high-temperature dryer cooling section. The time delay in transferring the hot corn and the fact that it is still moist at 18 to 20% moisture content will allow it to temper some before and during low-temperature drying. At this moisture level kernels do not have high internal stress levels that can lead to cracking and brittleness. This process will slowly cool and continue to dry the corn to about 14 or 15% moisture content, where the grain can be safely left in the bin for storage.

Equipment needed includes any high-temperature bin or self-contained dryer, low-temperature drying bins, and a well-organized conveying system. A grain cleaner and spreader are recommended for better airflow and more reliable low-temperature drying.

A combination system usually costs more than a single drying method, because storage bins are equipped for low-temperature drying including larger fans and supplemental heaters, and two drying methods and flexible grain handling are needed.

Combination drying gives both the high capacity and flexibility of high-temperature drying and the energy savings and superior grain quality of low-temperature drying. However, its operation is more challenging than dryeration or in-bin cooling.

Compared with conventional high-speed drying, combination drying can reduce energy use up to 50%, and high-temperature dryer capacity can be doubled or even tripled. Combination drying uses more electrical energy and less fuel, such as LP gas. Consider relative energy costs of electricity and LP gas to determine relative operating costs for both high-temperature and low-temperature drying. See Table 4-9 for a comparison of relative capacities and costs of high-temperature dryers with different cooling methods.

## RELATIVE COSTS AND CAPACITIES OF DRYING SYSTEMS

Capacity and operating costs of high-temperature drying depend in part on the grain cooling method: cooling in the dryer, in-bin cooling, dryeration, or combination high- and low-temperature drying. Table 4-8 shows a comparison of estimated relative operating costs.

Delayed cooling increases heated air dryer capacity, lowers dryer operating cost, and reduces stress cracks in grain. Delayed cooling increases the initial cost of drying, cooling, and handling equipment. Condensation is a problem of delayed cooling that must be considered and controlled.

**TABLE 4-8. Estimated relative operating costs of high-temperature drying with cooling methods.**

For drying/cooling corn from 25% to 15%.

| High-temperature drying with: | Relative capacity bu/unit of time | Gallons propane/100 bu | Electrical energy kWh/100 bu |
|---|---|---|---|
| In-dryer cooling | 1.0 | 20.0 | 10 |
| In-bin cooling | 1.35 | 17.5 | 8 |
| Dryeration | 1.6 | 14.5 | 7 |
| Combination drying | 2.5* | 8.0 | 70-110 |

*High-temperature drying capacity.

**EXAMPLE 4-5.**

Determine the relative capacities and operating costs if a high temperature dryer is converted to operate in full heat mode with delayed cooling. The high temperature dryer originally dries and cools 500 bu/hr. Assume energy costs are $1.70 per gallon for propane and 10¢/kWh for electricity.

**SOLUTION 4-5.**

**1. With in-dryer cooling**

The drying capacity is 500 bu/hr for a high-temperature dryer with in-dryer cooling. From Table 4-8, the high-temperature dryer uses 20 gallons of propane and 10 kWh of

▶ electricity/100 bu of corn dried from 25% to 15%. Dryer operating cost is:

(170¢/gal × 20 gal/100 bu) +
(10¢/kWh × 10 kWh/100 bu) =
34¢/bu + 1¢/bu = 35¢/bu

**2. With in-bin cooling**

From Table 4-9, the relative drying capacity is 1.35, and the high temperature dryer uses 17.5 gallons of propane and 8 kWh of electricity/100 bu of corn. Drying capacity increases to:

1.35 × 500 bu/hr = 675 bu/hr (+ 35%)

and operating costs decrease to:

(170¢/gal × 17.5 gal/100 bu) +
(10¢/kWh × 8 kWh/100 bu) =
29.8¢/bu + 0.8¢/bu = 30.6¢/bu (-13%)

**3. With dryeration**

From Table 4-8, the relative drying capacity is 1.6, and the high-temperature dryer uses 14.5 gallons of propane and 7 kWh of electricity/100 bu of corn. Drying capacity increases to:

1.6 × 500 bu/hr = 800 bu/hr (+ 60%)

and operating costs decrease to:

(170¢/gal × 14.5 gal/100 bu) +
(10¢/kWh × 7 kWh/100 bu) =
24.7 ¢/bu + 0.7 ¢/bu = 25.4 ¢/bu (-27%)

**4. With combination high/low temperature drying**

From Table 4-8, the relative drying capacity is 2.5, and the high-temperature dryer uses 8 gallons of propane and 90 kWh of electricity/100 bu of corn. Drying capacity increases to:

2.5 × 500 bu/hr = 1,250 bu/hr (+150%)

and operating costs decrease to:

(170¢/gal × 8 gal/100 bu) +
(10¢/kWh × 90 kWh/100 bu) =
13.6 ¢/bu + 9.0 ¢/bu = 22.6 ¢/bu (-35%)

These sample calculations indicate that drying capacities increase 35%, 60%, and 150%, and energy costs decrease 13%, 27%, and 35%, respectively, when a high temperature dryer is converted from operating in heat and cool mode (in-dryer cooling) to delayed in-bin cooling, dryeration or combination high/low temperature drying.

Results from Example 4-5 are summarized in Table 4-9. They include only drying capacity and energy costs. Note that for different propane and electricity prices, the energy cost will change, and thus the relative cost of drying and cooling will change. To compare drying systems, consider all equipment and labor costs. Compare equipment investments for wet holding, drying, cooling, and handling for each drying system. Drying equipment dealers and other farmers with similar equipment should have this information.

From investment cost, estimate annual fixed cost (depreciation, interest, repairs, and taxes) at 18% to 20% of initial equipment cost. Consult your Extension farm management specialist to help you evaluate the cost effectiveness of your planned system.

**TABLE 4-9. Summary of Example 4-5.**

Values are relative capacities and operating costs from Example 4-5 of operating a high temperature dryer with cooling hot corn in-dryer, in-bin cooling, dryeration, and combination high/low temperature dryer.

| High temperature drying, with: | Estimated drying capacity bu/hr | Estimated fuel operating cost ¢/bu |
|---|---|---|
| Cooling in dryer | 500 | 35.0 |
| In-bin cooling | 675 | 30.6 |
| Dryeration | 800 | 25.4 |
| Combination drying | 1,250* | 22.6 |

*High-temperature drying capacity.

CHAPTER 5

# GRAIN **STORAGE**

*Samuel McNeill, Ph.D., P.E., University of Kentucky*

Because grain quality and value do not improve during storage, the main goal of good storage management is to maintain grain quality after harvest by protecting it from damage caused by excess moisture or temperature, weather, rodents, insects, and mold growth (which can result in mycotoxin contamination). Generally, to maintain grain quality, all biological and insect activity must be controlled.

Several types of storage structures can be used successfully if managed properly. The needs of individual operations determine the type selected, such as permanent (metal or concrete) or temporary (sheds, warehouses, aerated outdoor piles, and bags). Regardless of the type used, proper selection, sizing, construction, and location are all essential to maintain grain value during storage.

## SELECTING GRAIN STORAGE

Management flexibility is related to the size and number of storage structures. If only a single large storage structure is available, flexibility is usually limited to a single crop and genetic trait, unless the storage is subdivided into separate compartments. Another factor that may affect flexibility is grain that needs to be stored a second year due to low prices or market demand; this could increase the risk of contaminating new grain due to possible insect and/or mold infestation problems. Additional factors that affect decisions about storage selection are crop production ratios (for example, corn to soybeans, wheat to grain sorghum, sunflower to barley) that may change or a given variety or attribute/trait that may show signs of field mold and need isolation.

A general rule from the past was to never have more than half of the total annual production of one grain type in a single storage structure. However, this may not apply to diversified operations where several different crops are held in one facility. In many cases, a mixture of storage sizes is best because different size structures allow more flexibility to meet changing needs and changing markets from year to year.

Grain is generally easier to manage in larger structures with smaller surface to volume ratios because the grain mass is less influenced by cooling and heating across a relatively smaller structural surface. This minimizes temperature gradient on an aerated grain mass and thus moisture movement.

To optimize grain storage costs, management flexibility, and production and marketing options, choose storage structure sizes large enough to meet your needs at a reasonable cost per bushel of stored grain.

### Round Metal Bins

Round metal bins made from corrugated steel are the most common on-farm grain storage structures. They are convenient to aerate, load, and unload. They are relatively low cost, compact, and fit well into a coordinated system. A wide range of grain drying, handling, aeration, and safety equipment is available to use with round bins. However, round bins have few alternate uses.

When selecting storage bins, there are no set rules for determining number and size. It is best to choose bins that best fit current needs and provide flexibility for the future. Select bins large enough to meet your storage needs at a reasonable cost per bushel. As bin size increases from 10,000 to 50,000 bushel, the cost per bushel drops sharply. For 50,000 to 100,000 bushel and larger bins, the cost per bushel of storage may still decrease, but not as rapidly.

The cost per bushel may even increase slightly with very large bins because of the substantial structural requirements for such things as foundations, reinforced sidewalls, trussed roofs, and additional construction labor. The cost per bushel difference between two 50,000-bushel bins and one 100,000-bushel bin may be greater, but this is a **one-time cost.** The flexibility of two smaller bins rather than one large bin can return the difference in cost because of more effective grain management and/or marketing.

Different handling equipment is needed for two smaller bins than for one large bin. Larger bins have greater diameter and height which may require a bucket elevator, distributor, and gravity spouts rather than portable and/or roof conveyors. Also, filling two smaller bins may require either a portable conveyor, a roof-top conveyor, or a bucket elevator with two gravity spouts. Unloading may require either moving a portable conveyor or installing a horizontal conveyor at ground level. However, the cost of this additional equipment is small when considered over the life of the bins, which is usually 20 years or longer.

Flat-bottom farm bins are usually filled only once per season. Round bins can be unloaded with underfloor augers emptying from an unload center sump on the floor. Plan on extra underfloor auger intakes for flexibility and convenience for larger bins. Consider an additional auger intake about 2 feet from the center sump and another near the bin door, as diagrammed in Figure 5-1. The intake 2 feet from the center can be used in an emergency if the center intake plugs. An intake near the door is convenient to move grain away from the door when unloading, thus providing for easy entry into the bin. Sweep augers minimize labor to remove all the grain from the bin, but they must be used with caution. Chapter 6, Safety, in this handbook provides more detail about safety issues involved in removing grain from a bin.

Fully perforated floors provide maximum flexibility in a totally integrated grain storage system. Perforated floors can be used for drying, cooling, and wet grain holding. They provide more uniform airflow at shallow depths but are more expensive than ducts or pads. Duct or pad air distribution systems are often selected because of lower costs. Regardless of the system chosen, a properly designed and installed duct or pad system can provide adequate air distribution.

Proper bin foundation and floor construction is important, especially for large bins. Follow manufacturers' specifications for constructing a concrete foundation. The bin floor should be level and have a slight grade away from the base ring of the outside wall to the edge of the foundation so water does not pool and/or freeze in this area. Any irregularities in the floor surface at or outside the bin wall that can collect water will cause serious problems. If bin size exceeds 50,000 bushels, follow the bin manufacturing guidelines for proper footing design which is a critical part of bin installation (3,000 pounds per square foot is typically a U.S. minimum standard for adequate soil bearing pressure). Drainage away from the bin is needed to prevent moisture buildup under the floor and around the bin. Figure 5-2 provides details for constructing the bin floor at least 12 inches above grade to minimize surface water problems and provide adequate clearance for the unloading auger.

Place a vapor retarder, usually at least a 6-mil plastic sheet on 4 to 6 inches of sand fill, before placing the concrete foundation. Seal the connection between the bin wall and concrete floor or foundation to prevent moisture and air from passing through. Inspect the bin before each

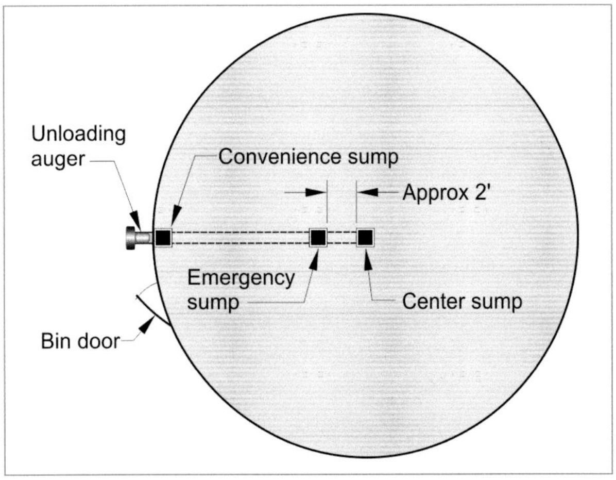


**FIGURE 5-1. Auger intake sumps in the floor of a round grain bin.**

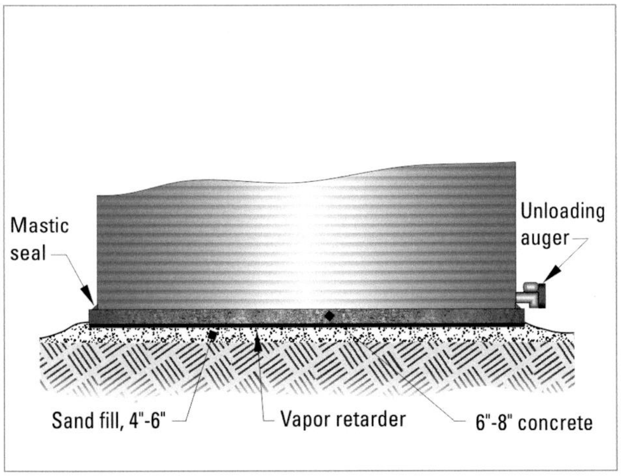


**FIGURE 5-2. Round bin wall and floor seal.** Construct the bin wall at least 12 inches above grade.

harvest to see if repairs are needed. Most commercial bin manufacturers provide a mastic seal, shown in Figure 5-2, between the steel bin wall and foundation. This pliable material provides sealing across irregularities in the concrete foundation.

Consider safety precautions and practices when installing round metal bins, include installing caged ladders, stairs, hand rails, and roof ladders on the outside of bins, following best practices, and keeping everyone outside the bin during loading and unloading.

Round metal farm bins are typically 27 to 60 feet in diameter, but can be smaller and as large as 135 feet in diameter. The volume capacity of round bins can be determined using the definition of a standard bushel, which is 1.245 $ft^3$. Table A-5 (and Spreadsheet S 5-1) gives bin capacities in bushels by volume based on a standard bushel for most bin diameters. The capacity of a level-filled round bin can be estimated using Equation 5-1.

**EQUATION 5-1.**

$$RBC = 0.785 \times D \times D \times H \times 0.803$$

RBC = round bin capacity, the estimated capacity of a level-filled round bin, bu
0.785 = constant
D = bin diameter, ft
H = bin height, ft
0.803 = conversion factor, $ft^3$ to bu, bu/$ft^3$

**For example:** A 30-foot diameter bin filled to a level depth of 18 feet holds an estimated 10,212 bushel (0.785 × 30 × 30 × 18 × 0.803). Values from Table A-5 may differ from the values found by using Equation 5-1 because of the approximate bu/$ft^3$ conversion factor. From Table A-5, a 30-foot diameter bin filled to a uniform depth of 18 feet contains 10,220 bushel by volume. However, bushel capacities by weight may differ from the values in Table A-5 because of compaction, test weight, and distribution of fines in the void space between the kernels. If each 1.245 $ft^3$ (one bushel by volume) of corn in the bin weighed 56 pounds (one bushel of U.S. No. 2 corn by weight at 15% moisture content), the bin would contain 10,220 bushel. Actual bin capacities are a somewhat greater than the computed volume due to grain compaction with increased grain depth. The only accurate method of determining the capacity of a bin is to weigh the grain loaded into the bin. Weight of grain loaded out after several months will likely be less due to handling and moisture shrink or dry matter loss. Other losses may occur from rodent, insect or mold activity.

A factor of 5% is commonly used to account for grain compaction. With a 5% compaction factor, a 30-foot diameter bin 18 feet level full of corn would contain 10,731 bushel (10,220 bu × 1.05) at 56 pound test weight. A higher or lower test weight would increase or decrease the capacity proportionately. A grain spreader tends to increase the compaction factor. A high percentage of fines can increase stored bulk weight because they are, by definition, smaller than grain or seed kernels, so the weight of fines increases the weight of stored grain by filling in the void space between kernels.

## Flat Storage Structures

Flat storage structures are commonly used for storing large quantities of dry grain—less so on-farm and more so at grain elevators. Usually, the storage is filled and unloaded only once a year because loading and unloading flat storage is not as convenient as with round bins. Flat grain storage offers the possibility of alternative uses, such as machinery and bale storage, after grain is removed. Or, portable bulkheads can be used to confine grain to specific areas in a building, so it can be used for more than one purpose at a time.

### Capacity

To estimate the capacity of a level-filled flat storage building use Equation 5-2 (or Spreadsheet S 5-2 by entering 0 for angle of repose).

**EQUATION 5-2.**

$$LFC = L \times W \times D \times 0.803$$

LFC = level-filled storage capacity, bu
L = storage length, ft
W = storage width, ft
D = grain depth at sidewall, ft
0.803 = conversion factor, $ft^3$ to bu, bu/$ft^3$

**For example:** A 100-foot long by 40-foot wide flat storage filled to a level depth of 4 feet holds an estimated 12,850 bushel (100 × 40 × 4 × 0.803).

Grain can be peaked to increase capacity in a flat storage building. Although peaked grain provides more storage capacity, properly checking and monitoring grain in a building is more difficult than in a typical round bin. Aeration cooling of peaked grain takes considerably longer than cooling level grain. However, leveling grain over a large surface is labor intensive and time consuming.

Most dry grains peak at a surface slope angle of 18 to 28 degrees, as detailed in Table 5-1. The actual filling angle depends on grain type, moisture content, amount of fines and foreign material in the grain, and height of drop during filling. Minimum and maximum values are shown in the table to illustrate the range of filling angles that can occur in a peaked grain mass. The actual peak height and storage capacity depends on the filling angle. The peak height can be estimated from Equation 5-3. For estimating grain capacity, use the average slope factor. However, use the maximum filling angle when evaluating the structural strength of a building used for grain storage.

### EQUATION 5-3.

$$PH = (0.5 \times W \times SF) + D$$

PH = height of peaked grain, ft
W = storage width, ft
SF = slope factor, Table 5-1.
D = grain depth at sidewall, ft

**For example:** The peaked grain height in a 40-foot wide building filled with soybeans or hard red spring wheat to a depth of 4 feet at the sidewall is 13.4 feet [(0.5 × 40 × 0.47) + 4]. From Table 5-1, the average filling angle is 25 degrees, and the slope factor is 0.47.

The grain volume in a typical, peaked flat storage building can be estimated by breaking it into three separate volumes, illustrated in Figure 5-3. Use Equations 5-4 to 5-7 to calculate the separate volumes. Multiply the total volume (V) by 0.803 bu/ft$^3$ to estimate the storage capacity in bushels. Alternatively, use Table A-7 (Spreadsheet S 5-2) to estimate the volume in bushels directly.

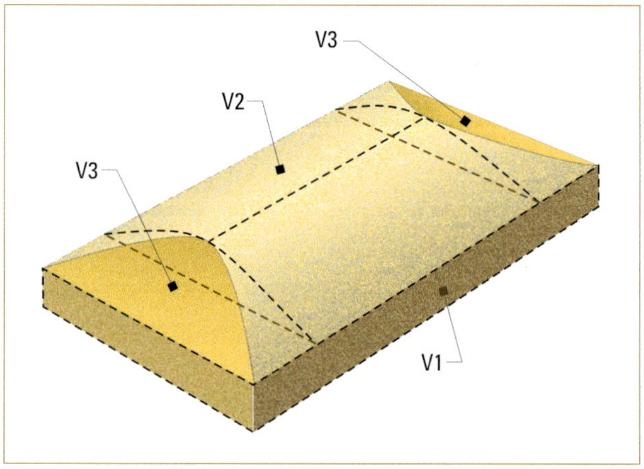


**FIGURE 5-3. The three grain volumes of a flat storage.**

### EQUATION 5-4.

$$V1 = W \times L \times D$$

**TABLE 5-1. Filling angles and corresponding slope factors.**

Filling angles are affected by grain moisture content, foreign material, and loading method and can vary from these values. When estimating storage capacity, use the average filling angle and slope factor.

| | Average | | Minimum | | Maximum | |
|---|---|---|---|---|---|---|
| Crop | Filling angle (°) | Slope Factor | Filling angle (°) | Slope Factor | Filling angle (°) | Slope Factor |
| Barley | 28 | 0.53 | 24 | 0.45 | 34 | 0.67 |
| Corn | 23 | 0.42 | 21 | 0.38 | 26 | 0.49 |
| Oats | 28 | 0.53 | 24 | 0.45 | 32 | 0.62 |
| Grain sorghum | 29 | 0.55 | 27 | 0.51 | 33 | 0.65 |
| Soybeans | 25 | 0.47 | 22 | 0.40 | 29 | 0.55 |
| Sunflower | | | | | | |
| • non-oil | 28 | 0.53 | 20 | 0.36 | 40 | 0.84 |
| • oil | 27 | 0.51 | 18 | 0.32 | 32 | 0.62 |
| Durum wheat | 23 | 0.42 | 22 | 0.40 | 25 | 0.47 |
| Hard red spring wheat | 25 | 0.47 | 19 | 0.34 | 38 | 0.78 |

**EQUATION 5-5.**

$$V2 = 0.25 \times W \times W \times (L - W) \times SF$$

**EQUATION 5-6.**

$$V3 = 0.065 \times W \times W \times W \times SF$$

**EQUATION 5-7.**

$$V = V1 + V2 + (2 \times V3)$$

V = total volume, $ft^3$
V1 = level fill volume, $ft^3$
V2 = middle triangular peak volume, $ft^3$
V3 = peak volume at one end, $ft^3$
W = building width, ft
L = building length, ft
D = grain depth at sidewall, ft
SF = slope factor, Table 5-1

**EXAMPLE 5-1.**

Calculate the volume of a 40-foot wide by 100-foot long flat grain storage building with corn 6 feet deep at the sidewalls and peaked at an average filling angle of 23 degrees, which is equal to a slope factor of 0.42 from Table 5-1.

**SOLUTION 5-1.**

$$V1 = 40 \times 100 \times 6 = 24{,}000\ ft^3$$

$$V2 = 0.25 \times 40 \times 40 \times (100 - 40) \times 0.42 = 10{,}080\ ft^3$$

$$V3 = 0.065 \times 40 \times 40 \times 40 \times 0.42 = 1{,}750\ ft^3$$

Total volume =
$$24{,}000 + 10{,}080 + (2 \times 1{,}750) = 37{,}580\ ft^3$$

Estimated storage capacity =
$$37{,}580\ ft^3 \times 0.803\ bu/ft^3 = 30{,}180\ bu$$

Estimated peak height =
$$(0.5 \times 40 \times 0.42) + 6 = 14.4\ ft$$

If grain is stored against the walls, they must be adequately reinforced to resist the large outward grain pressure. Structural failures not only cause building and grain losses, but can also cause injury and fatality. For information on proper wall reinforcement, read the section on Remodeling Post-Frame Buildings in this chapter.

### Managing stored grain

Flat storage structures can be used for NA/LT in-storage drying but to do so successfully must be level-filled. Special equipment is available to uniformly load a flat storage to achieve a level grain surface required for even drying. Drying air distribution ducting must be properly sized and placed to optimize airflow. Preferably grain held in flat storage should be thoroughly dried and cooled before it is placed in the structure. Moreover, quality losses will be minimized in flat storage by reducing grain moisture by 1 to 2% below levels considered safe for conventional storage.

Dry grain aeration in flat storage structures generally uses round or half-round metal or round plastic aeration ducts on the floor for air distribution. Mechanization of grain handling in flat storage is usually more difficult than in round grain bins. Grain is usually loaded into a flat storage with an overhead conveyor the length of the storage or by moving a portable inclined conveyor as the storage fills. Unloading is usually with portable augers, conveyors on the floor, a tractor with a bucket, a tractor-mounted inclined belt conveyor with a cross auger, a skid-steer bucket loader, or a pneumatic conveyor. Unloading usually must work around on-floor aeration ducts, which are removed as unloading progresses. Some structures have diagonal wall ties to the floor, which also interfere with grain unloading.

Before the concrete floor is poured in a flat storage building, a vapor retarder of 6- to 8-mil plastic sheeting should be laid to keep ground moisture from migrating or wicking through the floor into the grain. The floor of a flat storage building should be at least 6 inches above grade to minimize water problems. Drainage away from the storage is needed to prevent moisture buildup on the floor and around the storage building. Make sure the building site is well above 100-year flood plain elevations.

Rodents and birds can be difficult to control in flat storage building. Grain can be contaminated from mud or manure left in the storage by vehicles. Grease and hydraulic fluid on the floor also can contaminate grain. Thoroughly clean the building before storing grain, especially if fertilizer was stored in it previously. Do not store pesticides, which can be toxic, in buildings that will store grain. If in doubt, cover the concrete floor where grain is to be stored with 6-mil plastic sheeting.

*Consider these guidelines for successful grain storage in buildings:*

- Avoid use for drying. It is challenging to supply the high airflow rates required for timely NA/LT drying, so either let the grain dry in the field or use a heated air dryer, then cool the grain to ambient temperature BEFORE placing it in a flat storage structure. Place round or half-round perforated metal or round plastic aeration ducts on the floor. Consult the publication *Dry Grain Aeration Systems Design Handbook*, MWPS-29 for design criteria, sizing, and spacing. Use portable equipment for loading and unloading grain from inside the building.
- Use a combination of round bins and flat storage structures when practical. Loading the flat storage last and unloading it first minimizes flat storage time and maximizes its alternate uses. In some years, all or part of the building may be used for grain storage.
- Consider installing permanent grain loading and unloading equipment when planning to use flat storage structures for several years. Future expansion and improvement may include the building as a permanent storage structure.

### Location

Do not let the multiple use of a flat storage building interfere with the facility location such that it compromises good performance for either use. One use must dominate the location decision. If the primary purpose is grain storage, the importance of a good grain handling system design where the storage is loaded and unloaded through the grain center should be the primary consideration. If the building is used for occasional emergency grain storage, locate the building for maximum convenience for its most commonly used function, such as machinery storage and maintenance. Portable bulkheads or bins without roofs can be placed in buildings where sidewalls are not designed for grain loads.

## Existing Buildings and Structures

Many existing buildings and structures can be used to meet storage needs. Grain can be stored in some existing or temporary structures with relatively minor remodeling. However, converting structures such as a machine shed or cattle barn for grain storage usually requires major (and sometimes costly) remodeling.

Examples of small to moderate capacity buildings and structures that can be remodeled for grain storage include corn cribs, empty livestock buildings, machinery sheds, barns, upright silos, and horizontal silos. Other temporary storage ideas include stacked hay bales lined with plastic as temporary sidewalls to increase pile depth and volume in areas such as concrete driveways; portable bulkheads; round plastic silage bags; and covered, aerated outdoor piles. These can be used successfully, provided proper grain management practices, including grain aeration, are implemented. To prevent structural failures, it is essential that all existing buildings and temporary structures be adequately reinforced to withstand the expected grain pressures.

### Remodeling considerations

Grain in a building exerts considerable and often additional pressure on the outside walls, so the building frame and lining material must withstand large horizontal loads that in most cases exceed the original building's structural design. Fastenings, anchorage, and the foundation must withstand these loads without breaking or excessive movement. Before storing grain in an existing building, determine the strength of the walls, floor, and other structural members to be sure they can withstand the expected pressures. Structural failures not only cause serious structural damage and grain losses, but can also cause serious personal injury or fatality. Consult your state extension agricultural engineer or a professional structural engineer who is experienced in grain storage design to evaluate existing buildings and structures for grain storage.

Shallow grain piles cost less and are easier to contain than deep ones. For example, the total horizontal wall pressure from a 10-foot deep sidewall depth is more than six times greater than the pressure from a 4-foot sidewall depth. The horizontal pressure exerted by a peaked pile of grain on walls is about 30% greater than the pressure from a level filled pile of grain even though the grain depths are the same at the wall.

Lay a plastic vapor barrier between the floor and grain to prevent moisture from migrating through the floor into the grain. If a vapor barrier was installed under the floor, a second vapor barrier is not needed. Check the roofing and flashing at hatches and vents for water-tightness. Reduce bird and rodent damage by screening all openings. Minimize ledges and cracks that hold or trap old grain because they are ideal for harboring insects. Maintain a rodent bait program. Consult an extension entomologist or an entomologist familiar with stored grain insects about approved insecticide treatment choices. Work with an extension agricultural engineer or a consultant to gather information about designing an aeration system and for operating guidelines.

## Upright silos for storing dry grain

Upright, roofed silos of any type can store dry grain. This includes silos made from concrete and clay tile with external steel hoops, cast concrete, metal, fiberglass, and plastic.

*Take the following precautions when using upright silos for dry grain storage:*

- Be sure the silo is structurally sound. Silos designed solely for silage are usually not structurally adequate for storing dry grain. Some silo manufacturers may specify a depth limitation for grain, such as 40 to 50% of the depth used for silage. Check with the structure's original manufacturer, a professional structural engineer, or your state extension agricultural engineer for recommended structural reinforcing. Avoid old silos, especially if reinforcing steel is imbedded in the wall. Older silos with weathered walls may not be able to withstand the pressure exerted by the grain and may collapse. Walls must be sound so that moisture does not enter and spoil the grain.
- Keep the maximum dry-grain storage depth to no more than 50 feet in concrete stave silos. Dry grain puts a higher load on the walls which could cause the staves to break. Before storing grain deeper than 50 feet, consult your state extension agricultural engineer or a professional structural engineer to determine the structural adequacy of the silo.

Use Table A-8 (Spreadsheet S 5-3) to calculate silo capacities for dry grain.

- Aerate grain in silos to prevent convection currents and moisture migration. Use positive pressure aeration for silos adapted for dry grain. Figure 5-4 shows a suggested aeration system for oxygen-limiting silos, while Figure 5-5 shows an aeration system for concrete stave silos.
- Load at the center of the silo to prevent off-center loading that could cause structural damage and result in a structural failure and/or collapse, which could injure or kill workers or livestock.
- Unload the silo from the bottom center—do not draw from the side because unbalanced loads can cause structural damage or failure. Use a portable horizontal auger to unload grain through the aeration duct, pulling grain from the center of the silo so wall loads remain uniform.
- Be sure silo roofs and stave joints do not leak during blowing rains. Consider using a spray-on concrete sealing compound to seal the inside wall. Allow time for proper curing before filling with dry grain.
- Install some type of floor vapor retarder. This can be a concrete floor or a layer of sand at least 6 inches deep with a plastic vapor retarder. If you expect to use the upright silo for high moisture grain or silage in the future, a sand floor provides more flexibility than concrete.

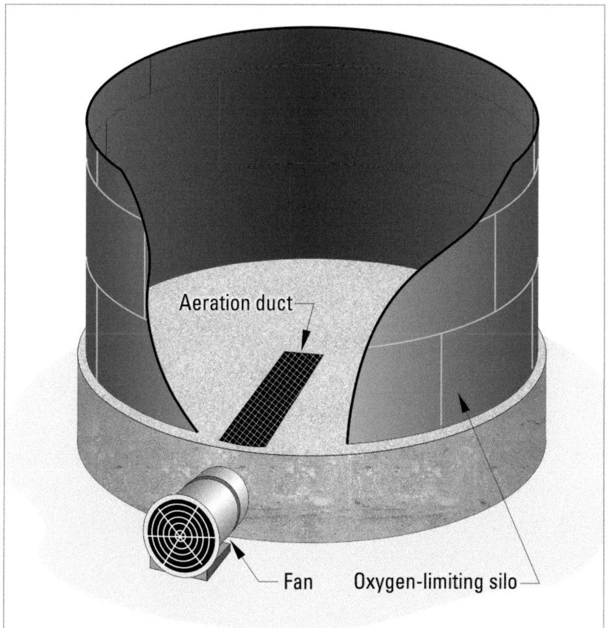


**FIGURE 5-4. Aeration duct and fan in an oxygen-limiting silo converted for dry grain storage.**

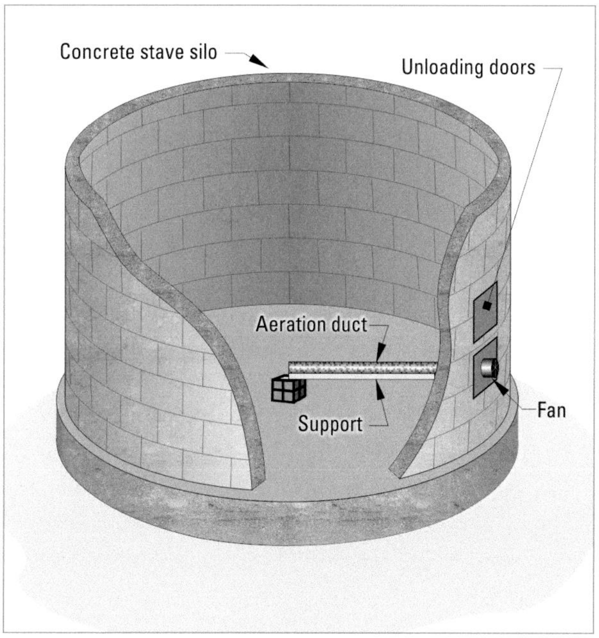


**FIGURE 5-5. Aeration duct and fan in a concrete stave silo converted for dry grain storage.**
Unload through the aeration duct with a portable auger.

### Hopper-bottom bins

Hopper-bottom bins with 45-degree hopper slopes allow gravity unloading and are self-cleaning. Hopper-bottom bins are more expensive than flat-bottom bins, so use them only where they will be filled and emptied frequently, such as in wet grain holding, overhead loadout bins, wet holding above a high temperature automatic batch or continuous flow dryer, in dryeration systems or feed processing. An elevated hopper-bottom bin requires considerable structural support, which makes it even more expensive. Most hopper-bottom storages are commercially available round corrugated steel units. However, they can also be square steel or small wooden bins built with steel, lumber or concrete supports.

### Outdoor piles

Large, covered outdoor piles are often used for emergency/temporary storage and work best when temperatures are below 50°F. As with other flat storage options, large diameter corrugated metal rings (much like a grain bin) or specially designed retaining wall sections will add height and considerable capacity to the structure and lower the storage cost per bushel. Use Table A-10 (Spreadsheet S 5-4) to estimate storage capacities of round and oval-shaped outdoor piles, shown in Figure 5-6. Grain losses in piles are usually higher than in conventional storages, so proper grain management is essential to minimize losses. In particular, the site should be well drained, have a packed base of sand or small gravel with an underlying heavy plastic sheet and overlaid with a geotextile or impervious material to prevent contamination of the grain. Moreover, grain should be dried 1 or 2 percentage points lower than conventional storage and be reasonably clean. Pockets of wet grain or foreign material can cause spoilage. Use a negative-pressure aeration system to hold the cover in place, and check for damage from birds and rodents at least weekly. Additional tarp anchorage to the sidewall, cables secured across the pile, or equally spaced weights (such as old tires) are recommended on the pile cover in case the suction fan motors fail or a power outage occurs. Major grain losses have occurred when high winds damaged grain pile covers that were not secured adequately. Moisture due to rain and snow can also infiltrate through tears in the pile cover or broken seams. Grain monitoring equipment is especially important and should be used to check the temperature and carbon dioxide concentrations of air exhausted from the aeration fans. Contact your state extension agricultural engineer for additional details for storing grain in covered, aerated outdoor piles.

### Large silo bags for dry grain

Polyethylene bags of adequate thickness and strength (similar in size and function to silage bags but designed for grain storage) can also be used successfully for temporary storage of dry grain. The same guidelines mentioned earlier for other temporary structures apply to grain temperature and moisture target levels (below 50°F and 1 or 2 percentage points lower moisture than conventional storage). They are often used on rented farms and in remote or isolated fields with limited, secure access. They are designed for airtight (hermetic) storage, which means that the normal respiration of grain and/or insect and mold activity consumes oxygen and increases carbon dioxide, which in turn preserves the grain as long as the silo bag remains sealed unless heavily infested by insect or presence of substantial mold spoilage,

**FIGURE 5-6. Large covered and aerated outdoor piles can be used for short-term storage preferably when outdoor temperatures are below 50°F.**

carbon dioxide levels will remain well below 10% during storage of dry grain. If the seal is broken by birds, rodents, or other animals, and moisture enters, spoilage can occur in that area of the bag. Thus, bags should be checked weekly for damage.

Large silo bags, such as the example in Figure 5-7, are typically 8 or 10 feet in diameter, up to 500 feet in length, and able to store up to 30,000 bushels per bag. Use Table A-11 (Spreadsheet S 5-5) to estimate capacities. Specialized handling equipment is needed to load and unload the bags. This equipment can be rented in most areas, so bags can be used quickly and inexpensively to provide storage as needed when harvest exceeds available storage capacity. Silo bags require a larger footprint (more land area) than most any other storage option, so the cost per unit of area can be quite high, but land space is usually not a constraint in rural areas. Bags should be placed with the length in a north-south direction to balance exposure to the sun and minimize moisture condensation and migration. When bags are placed in the field, the site should be prepared by flattening corn, wheat, or soybean stubble to minimize punctures when setting up and filling the bags. A few passes with a track vehicle could make this ground surface preparation task go quickly. Empty bags weigh as much as 300 pounds, and are spooled on the filling unit, therefore minimize labor before filling. They are cut open when unloaded and cannot be reused for grain storage, so they should be recycled wherever possible.

**FIGURE 5-7. Grain bags can be used for temporary storage but must be managed well to protect them from damage by birds, rodents, deer or other animals.**
*Courtesy of Loftness Specialized Equipment.*

# REMODELING POST-FRAME BUILDINGS

For purposes of discussion in this handbook, a post-frame building can consist of either square posts, or round poles made from wood or steel. These remodeling ideas can be used for either. Provide appropriate personal protection equipment (PPE) to all workers when remodeling buildings for grain storage.

To store grain in a post-frame building, extra posts may be needed, depending on grain depth and the building's design. Consult your state extension agricultural engineer or a professional structural engineer experienced in grain storage design to evaluate existing buildings and structures for grain storage for recommended techniques to properly reinforce walls.

The following sections present some remodeling ideas for storing grain up to 4 feet deep in post-frame buildings. Follow design specifications and construction details exactly to eliminate the chance of a building failure. The following design ideas are for ground-level storages only. Before adding grain storage to a hay loft floor or other elevated floor, consult a professional engineer experienced in structural design for grain storage facilities.

Contamination from wood preservatives may become a problem in post-frame buildings. Federal regulations restrict use of preservatives in contact with grain and feed. Do not use creosote or pentachlorophenol where feed or grain is stored.

## Four-foot High Plywood Liner

Lining the lower 4 feet of an existing post-frame building with plywood sheet to retain grain is fairly economical and does not interfere with future uses of the building. A paved or concrete floor makes unloading grain easier.

With 6 × 6-inch posts or larger, 8 feet on center, 4 feet in the ground and no more than 14 feet above the ground, grain can be safely filled up to 4 feet deep at the plywood-lined sidewalls. For building frames with posts smaller than 6 × 6 inches, consult your state extension agricultural engineer or a professional structural engineer experienced in grain storage design. Figure 5-8a shows construction details for adding a 4-foot high plywood wall between the posts to contain the grain. Cut a 2 × 6 inch sill and plate to fit between the posts. Assemble the frame with 16d nails. Attach the 2 × 6 inch sill and 2 × 6 inch plate to the 2 × 4 inch studs with two 16d nails at each end. Tip the frame into place. Fasten the sills and plates to the posts with steel framing anchors attached with #8 × 1.5 inch galvanized hex head lag screws. Line the wall with 4 × 8 feet by 5/8- or 3/4-inch exterior grade plywood sheets. Select plywood with an Identification Index of 42/20 Ext. Alternatively, use 5/8-inch CDX plywood if from Group 1 or 2 species and 3/4-inch plywood if from Group 3, 4, or 5 species. Framing lumber should be No. 2 or better, Southern Pine or equivalent.

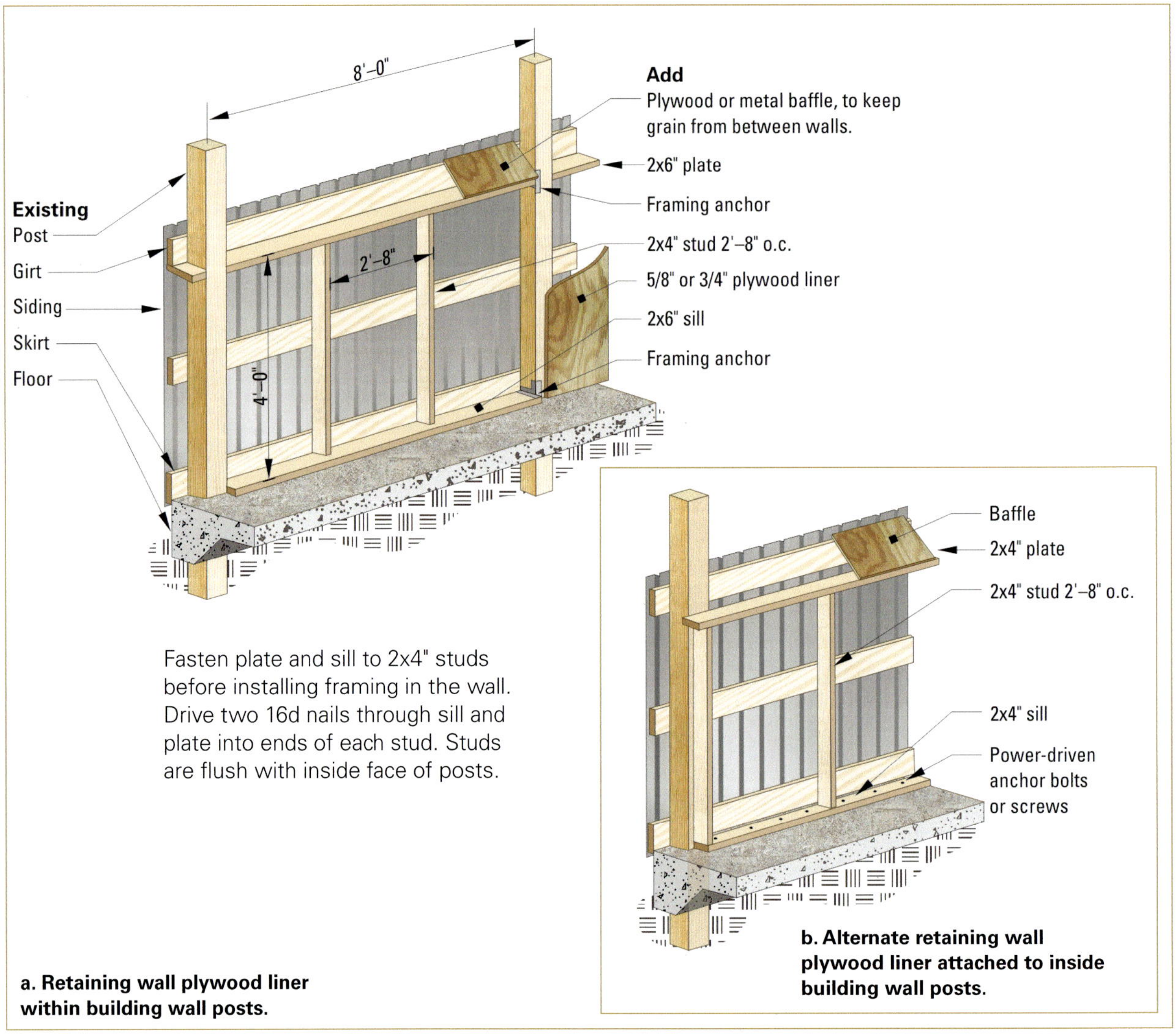


**a. Retaining wall plywood liner within building wall posts.**

**b. Alternate retaining wall plywood liner attached to inside building wall posts.**

**FIGURE 5-8. Retaining wall plywood liner for post-frame buildings.**

To keep grain from falling between the support wall and building wall, add a plywood or metal baffle to the top of the support wall along a wall girt. Do not pile grain deeper than 4 feet against the support wall.

An alternative to the construction shown in Figure 5-8b is to attach the plywood liner wall to the inside building wall posts. The sill and plate can be continuous across one or more posts; add a stud at each post to support the plywood sheets. Anchor the sill to the floor with anchor screws or with anchor nails driven through the sill with a power tool.

## Four-foot High Concrete Wall

A 4-foot high concrete wall can be cast between posts of a new or existing post-frame building to reinforce it for grain retaining. Figure 5-9 illustrates this practice. Concrete walls can be cast between posts up to 8 feet on center. When constructing a new building, tie the wall and floor together

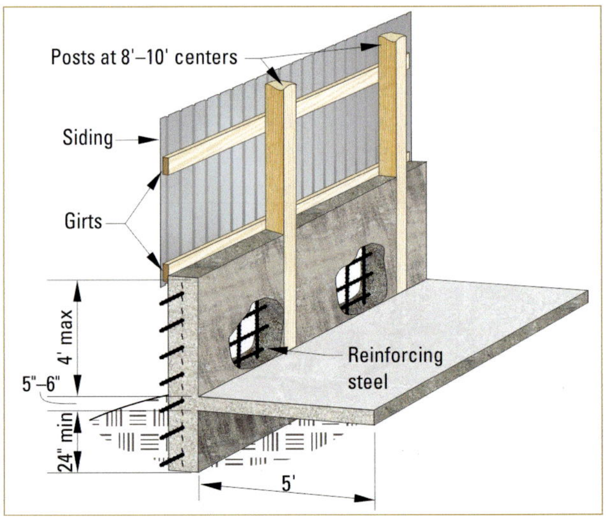


**FIGURE 5-9. Concrete wall to retain grain between posts of post-frame buildings.**

with reinforcing steel (rebar) rods. In new construction, cast the wall at least 24 inches below grade. To add a concrete wall to existing buildings with floors, cast the wall between posts on top of the floor and seal the joint with mastic.

For specific designs, construction techniques, and detailed information about concrete quality, steel rebar size, and spacing required for farm buildings, and other information about concrete placement, consult *Farm and Home Concrete,* MWPS-35. This handbook is available from MWPS at www.mpws.org.

## Movable Grain Walls

Movable grain walls, shown in Figure 5-10, can be built of dimensioned lumber and plywood or rough-cut lumber of equal structural strength. These walls are self-supporting and can convert all or part of a building to retain grain for storage, as shown in Figure 5-11a.

**FIGURE 5-10. Movable grain walls with a sloped floor.**

## Bin Rings

Round, metal bin rings, seen in Figure 5-11b, can be conveniently and economically erected inside an existing building to provide emergency grain storage. Set the base ring on the existing floor and anchor as recommended by the manufacturer. Add two or three additional rings where possible to add capacity without additional remodeling to the building walls. Install an unloading auger in the center of the rings to remove center core and for emptying most of the grain.

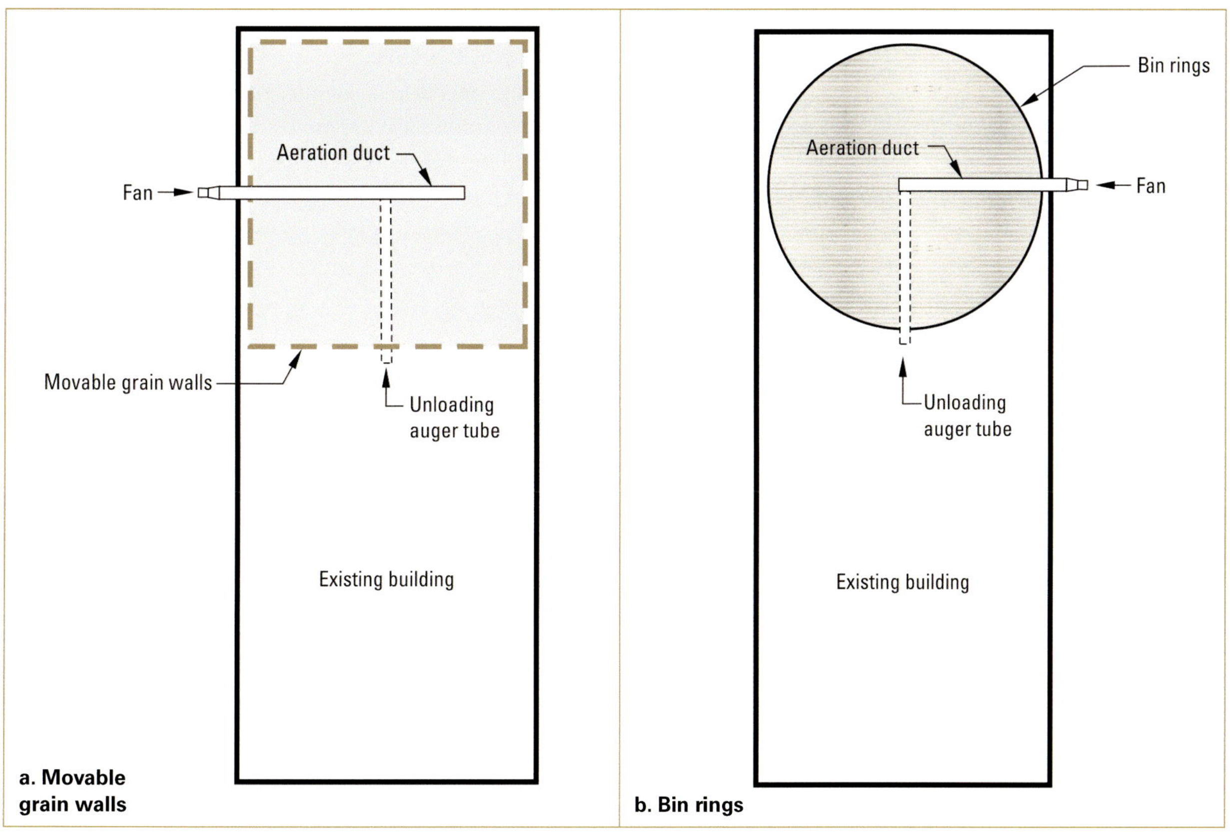


**FIGURE 5-11. Options for retaining grain for storage in existing buildings with movable grain walls (a) or bin rings (b).**

Using grain-bin sidewall rings as an alternative to wall reinforcement may leave building corners that are difficult to utilize, but round rings stacked higher can hold more grain than square facilities limited to 4 feet high sidewall liners. The round rings can be managed much more efficiently with minimal remodeling.

**For example:** A 42-foot square pile which is 4 feet deep at the walls and peaked at 23 degrees (12.9 feet deep in the center) holds 8,975 bushel. By comparison, three, 44-inch wide round bin rings, 42 feet in diameter with an 11-foot sidewall in the same building with level grain will hold 12,240 bushel.

# MANAGING GRAIN IN STORAGE

Properly managing grain in storage is important to maintain grain quality and quantity after harvest. Putting grain in storage does not improve its quality, but with good management, quality should not degrade. Generally, to maintain quality, grain moisture and grain temperature as well as all biological activity (molds, insects) must be controlled.

*Factors that influence whether grain goes out of condition or degrades in quality are the following:*

- Grain moisture content
- Grain temperature
- Initial grain quality
- Amount and distribution of broken grain fines and foreign material
- Presence of insects, molds, rodents and birds

Maximum recommended storage times for clean, aerated shelled corn at different moisture and temperature levels are shown in Table 5-2. The table lists the time corn may be stored before a 0.5% reduction in dry matter is expected to occur, which usually results in a loss of one U.S. Grade level. Allowable storage times for corn that is immature or has a lot of damage from insect or mold activity in the field, or from mechanical damage during harvest or handling should be reduced by 30 to 50% for the same temperature and moisture values. Note that storage times can be doubled as moisture and temperature levels are reduced. It is important to keep in mind that there is a cumulative effect on storability.

**For example:** If 18% moisture corn is held in aerated storage for 30 days at 60°F, one-half of the allowable storage life is used. When this corn is cooled to 50°F, the recommended maximum storage time is only 65 days.

## Moisture Content

The average moisture content of all grain in storage does not determine storability. Spoilage can occur in isolated locations or pockets where grain moisture is high due to insufficient drying or moisture accumulation due to condensation or infiltration. Recall that Table 4-1 gives the maximum recommended moisture content for several grains and oilseeds based on the geographical location and storage period. These moisture levels are based on the conditions inside the grain mass for the average ambient temperature that is expected during storage. These conditions set the relative humidity (RH) of the air (void) space between kernels, which should be at or below 65% to maintain an environment that is sufficiently dry to prevent mold activity.

Figure 5-12 shows the maximum safe moisture content of different grains and oilseeds for a range of grain temperatures at a relative humidity of 65% within the grain mass. Conditions above these lines are at risk from mold activity (RH > 65%), whereas conditions below these lines have reduced risk of spoilage. Keep in mind that once grain is uniformly dried for the intended storage period, it should be monitored regularly to be sure no water leaks into the structure and that no other causes contribute to an increase in the moisture content anywhere in the grain mass.

## Grain Temperature

Grain temperature also must be controlled to minimize spoilage. As previously discussed, lower grain moisture and temperature levels decrease biological (mostly molds and insects) activity, thereby increasing safe storage periods.

Grain is a good insulator, so it does not cool uniformly throughout the bulk as outside temperatures drop during late fall and winter. Grain near the bin wall cools, but

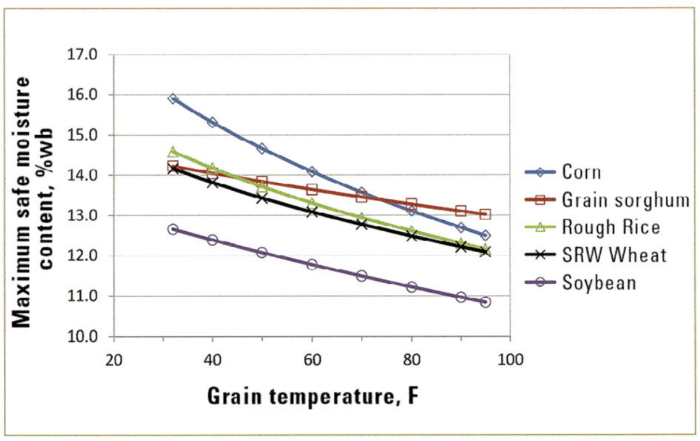


**FIGURE 5-12. Maximum safe moisture contents for selected grains and oilseeds over a range of temperature levels, based on a relative humidity of 65%.**

*Source: American Society of Agricultural and Biological Engineers.*

**TABLE 5-2. Maximum storage life for shelled corn.**

Values are for corn held at a constant temperature, that is, with respiration heat removed by aeration. These values are the times estimated for shelled corn held at various temperature-moisture combinations to reach 0.5% dry matter loss, known as the acceptable storage life. Shelled corn would drop by one U.S. Grade level on average at 0.5% dry mater loss – from U.S. No. 2 to U.S. No. 3. Such grain is usually visibly moldy. Do not use all the storage time on your farm; save some storage time for the next users so they have a quality product to work with. Therefore, it is advisable to be conservative when using the allowable storage life. Example: 26% moisture content (m.c.) corn held at 60°F for 4 days uses 1/2 of its storage life (4 days/8 days). If then dried to 14% m.c. and held at 50°F, the corn has only 650 days remaining (1/2 × 1,300).

*NOTE: Table 5-2 was generated assuming a low normal grain damage level. Damage level to kernels incurred during combining at harvest has a huge influence on the allowable storage life. Fungi grow faster on kernels with high levels of damage, which could decrease the allowable storage life by half of those estimated in the table. Other factors such as hybrid type can influence the allowable storage time. Thus, using a conservative estimate is important. Adopted from ASABE Standard D535 (ASABE, 2010).*

| | Corn moisture, % wet basis | | | | | | | | | | | | |
|---|---|---|---|---|---|---|---|---|---|---|---|---|---|
| **°F** | **°C** | **14** | **15** | **16** | **18** | **20** | **22** | **24** | **26** | **28** | **30** | **32** | **34** |
| 35 | 1.7 | 4,374 | 2,093 | 1,129 | 431 | 213 | 126 | 85 | 63 | 49 | 41 | 35 | 31 |
| 40 | 4.4 | 2,919 | 1,396 | 754 | 288 | 142 | 84 | 57 | 42 | 33 | 27 | 23 | 20 |
| 45 | 7.2 | 1,948 | 932 | 503 | 192 | 95 | 56 | 38 | 28 | 22 | 18 | 16 | 14 |
| 50 | 10 | 1,300 | 622 | 336 | 128 | 63 | 38 | 25 | 19 | 15 | 12 | 10 | 9 |
| 55 | 12.8 | 867 | 415 | 224 | 85 | 42 | 25 | 17 | 12 | 10 | 8 | 7 | 6 |
| 60 | 15.6 | 579 | 277 | 149 | 57 | 28 | 17 | 11 | 8 | 7 | 5 | 5 | 4 |
| 65 | 18.3 | 431 | 206 | 111 | 42 | 21 | 13 | 9 | 7 | 5 | 4 | 4 | 3 |
| 70 | 21.1 | 322 | 154 | 83 | 32 | 16 | 10 | 7 | 5 | 4 | 4 | 3 | 3 |
| 75 | 23.9 | 241 | 115 | 62 | 24 | 12 | 8 | 5 | 4 | 3 | 3 | 2 | 2 |
| 80 | 26.7 | 180 | 86 | 47 | 18 | 9 | 6 | 4 | 3 | 3 | 2 | 2 | 2 |
| 85 | 29.4 | 135 | 65 | 35 | 13 | 7 | 5 | 4 | 3 | 3 | 2 | 2 | 2 |
| 90 | 32.2 | 101 | 48 | 26 | 10 | 6 | 4 | 3 | 3 | 3 | 2 | 2 | 2 |
| 95 | 35.0 | 76 | 36 | 20 | 7 | 4 | 4 | 3 | 3 | 2 | 2 | 2 | 1 |
| 100 | 37.8 | 57 | 27 | 15 | 6 | 4 | 3 | 3 | 2 | 2 | 2 | 2 | 1 |
| 105 | 40.6 | 42 | 20 | 11 | 4 | 3 | 3 | 2 | 2 | 2 | 2 | 2 | 1 |
| 110 | 43.3 | 32 | 15 | 8 | 3 | 2 | 2 | 2 | 2 | 2 | 2 | 2 | 1 |
| 115 | 46.1 | 24 | 11 | 6 | 2 | 2 | 2 | 2 | 2 | 2 | 2 | 2 | 1 |
| 120 | 48.9 | 18 | 8 | 5 | 2 | 2 | 2 | 2 | 2 | 2 | 2 | 2 | 1 |

[a]D = 30%, MD = MH = MF = 1, Times calculated using °F temperature values.

grain temperatures near the center remain unchanged if not aerated. Conversely, once grain is cooled with aeration to the desired final storage temperature during the fall and fans are sealed, most of the grain remains cool, as illustrated in Figure 5-13.

## Aeration Systems

Aeration is not intended for drying grain but is essential for successful dry grain storage. However, each time grain is aerated, some grain moisture is removed (about 0.25 to 0.3 percentage points of moisture per 10°F temperature decrease during an aeration cycle), so some market weight is lost (shrink loss). A properly managed aeration system moves air through grain to lower grain temperatures throughout the mass and reduce biological (molds, insects) activity. Most insects enter grain bins in the top or headspace, especially if base openings such as unload augers and aeration fans are properly sealed. Before aerating, core the grain mass by unloading a portion of grain in order to level or slightly inverse the grain surface. That assures uniform airflow across the diameter of the bin.

Aerate grain during the fall and early winter to uniformly cool the grain mass from bottom to top below 35 to 40°F. It may take two to three aeration cycles in October (to below 45°F), November (to below 40°F) and December (to below 35°F) to achieve the target temperatures. In the northern parts of the Midwest stored grain can be cooled below 30 to 35°F because of colder average winter temperatures. In all locations, seal aeration fans after each aeration cycle is complete to minimize convection air currents. It is especially important to seal the aeration fan(s) after the last cooling cycle to hold temperatures as low as possible along the floor during daily and seasonal temperature changes throughout the winter and into the spring and summer.

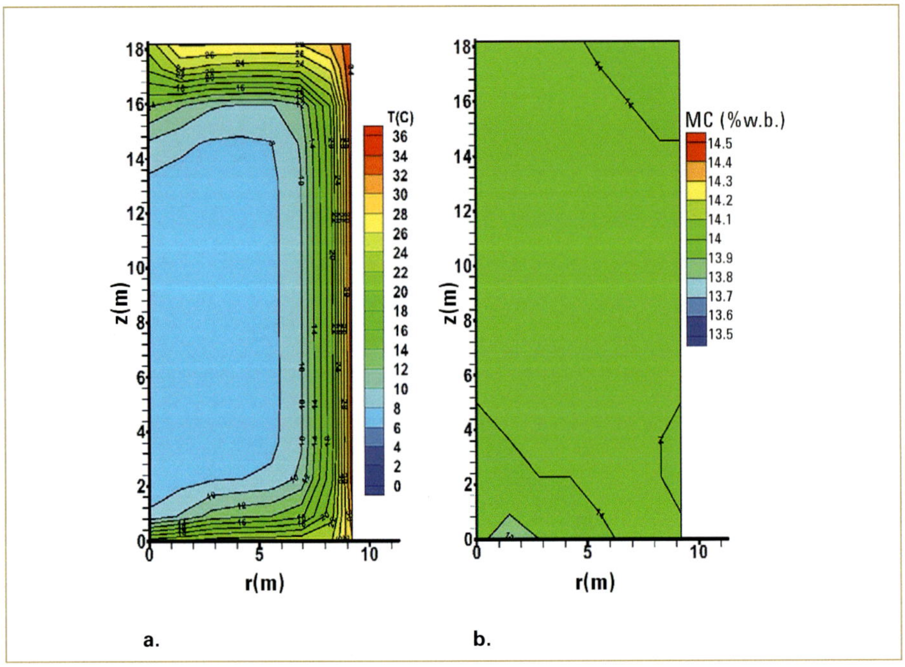


**FIGURE 5-13. Predicted temperature (a) and moisture (b) changes after 10 months of storage in Indiana of dry, cool corn that was cooled by aeration during the previous fall and the fan sealed after aeration.(60- x 60-foot bin with 135,000 bushel).** Note that grain temperatures in the center of the bin remain stable (a), and grain near the perimeter warms to near the average monthly temperature. However, most of the grain remains below 60°F, and the moisture content throughout the bin is relatively unchanged (b).

In tall concrete silos and for the storage of wheat harvested in the heat of the summer at 90 to 110°F, suction cooling can greatly reduce insect activity in the upper portion of the grain mass. At night time cooler ambient air can be drawn into the headspace of the silo and drawn into the upper grain layers to reduce wheat temperatures. Suction fans must be stopped during the day time to avoid drawing hot ambient air that can additionally be warmed due to the solar heating of roofs to counter the effect of night time cooling.

High static pressures that cause warming of cooling air across aeration fans and the length of time it takes to move an aeration front through a tall wheat silo at 0.05 cfm/bu (i.e., at least 300 hours) makes suction cooling for the reduction of market activity in the upper grain layer effective.

Insects are most active from 75 to 90°F. Entomologists have found that by reducing grain temperatures by 10°F (such as 90 to 80°F) reduces insect reproductive activity substantially, and cooling further to below 60°F stops reproductive activity altogether. This is one application where downdraft (suction) aeration has an advantage over updraft (push) aeration. In all other applications updraft aeration is preferred especially in the Midwestern U.S. where frozen roof exhausters and vents could collapse a bin roof if downdraft aeration fans were turned on during such winter conditions

Proper operating and management practices must be followed with any aeration system. Even the best designed aeration system can fail if not operated correctly. A modestly designed system with proper management is better than an elaborate system with poor management. For more information about proper operation of grain aeration systems, refer to MWPS publication AED-20, *Managing Dry Grain in Storage*.

### Airflow rates

Grain can be successfully aerated with airflow rates as low as 0.05 (1/20) cfm/bu to over 1 cfm/bu, which is a common rate for bin dryers. Although drying fans can be used for aeration, low capacity aeration fans cannot be used for drying because they do not move enough air to dry the grain before spoilage occurs.

Airflow rate determines the time required to complete one aeration cycle. An aeration cycle is the time it takes the temperature front to pass through the grain mass to change the temperature of all grain in the bin.

> **For example:** At the common aeration airflow rate of 0.1 cfm/bu (cubic feet per minute per bushel), it requires about 150 hours to change the temperature of all corn (or other grains with similar test weight) in the bin. For grains and oilseeds with other test weights, refer to Equation 1.3 in *Dry Grain Aeration Systems Design Handbook,* MWPS-29.

To be sure of complete cooling, you must monitor the change in grain temperature. For a negative pressure or suction aeration system, check the temperature at the fan discharge. With positive pressure systems, where air is blown up through the grain, check grain temperatures at the grain surface. A lower change in grain temperature over 10 to 15 minutes indicates that the cooling front has passed through the grain. Ideally, install temperature cables to monitor movement of aeration cooling fronts through the grain mass conveniently and reliably.

Updraft aeration fans compress the air, adding mechanical heat-of-compression to ambient air temperature. Heat off the fan motor from direct drive fans also increases ambient air temperatures by 0.8 to 1.5°F. As static pressure increases, heat-of-compression temperatures increase. This temperature rise can be measured easily by inserting a thermometer through a small hole in the air duct between the fan and bin plenum. Temperature increases of 3 to 5°F are not uncommon for aeration systems that push air through deep grain bins or small grains such as wheat and grain sorghum (about 0.8°F per 1 inch of water column/static pressure). In cases where static pressures are high (5 to 8 inches water column) systems only cools grain to within 4 to 6°F of the ambient temperature. Suction aeration systems can cool grain closer to the ambient temperature as long as solar radiation does not heat the headspace air too high before it is drawn into the upper grain layer.

Higher airflow rates decrease cooling time proportionately. For example, if 0.1 cfm/bu cools grain in about 150 hours of fan time, then 0.2 cfm/bu cools grain in about 75 hours of fan runtime, and 0.5 cfm/bu in about 30 hours. Cooling time is essentially independent of the amount of grain temperature change and the grain depth.

Keep in mind that airflow rates are different for different grains. Small, elongated kernels like wheat, barley, and rice create small void spaces between kernels, whereas large, more rounded kernels like corn and soybeans have larger void spaces. Also, clean smaller kernels offer more resistance to airflow (as measured by static pressure against the fan) than clean larger kernels. The FANS on-line program (http://webapps.bbe.umn.edu/fans/) can be used to determine the airflow rate through different grains for a given bin-fan(s) combination. Any size bin and fan that is commercially available in the U.S. can be evaluated for the primary grain crops.

**For example:** A 48-ft diameter bin with in-floor aeration ducts and a sidewall height of 32 feet will hold approximately 46,300 bushel when the surface is level. A typical 18-in diameter axial fan with a 1.5-horsepower motor will provide 0.09 cfm/bu for corn, but only 0.06 cfm/bu when the bin is full of wheat, as detailed in Table 5-3. The time required to move a cooling front through each crop when filled to this depth would be 167 and 250 hours of fan operation for corn and wheat, respectively, for this fan. If shorter times are desired, the depth should be reduced accordingly for this fan. For example, when wheat is filled to a depth of 24 feet in this bin, the airflow rate would be 0.09 cfm/bu, resulting in a cooling front that would require the same number of fan run hours (i.e., 167 hours).

## Air distribution

Proper design and arrangement of an aeration system is important for proper air distribution. In round bins, full perforated floors are generally used to provide more even

**TABLE 5-3. Predicted airflow rates for corn and wheat in a 48-foot diameter and 32-foot eave height round bin with a 1.5 hp axial fan (18-inch diameter).**

| | | Corn | | | Wheat | | |
|---|---|---|---|---|---|---|---|
| Grain depth ft | Grain volume bu | Static pressure (inches of water) | Airflow cfm | Airflow rate cfm/bu | Static pressure (inches of water) | Airflow cfm | Airflow rate cfm/bu |
| 3 | 4,343 | 0.56 | 4,576 | 1.05 | 0.72 | 4,366 | 1.01 |
| 6 | 8,686 | 0.61 | 4,504 | 0.52 | 0.91 | 4,116 | 0.47 |
| 9 | 13,029 | 0.67 | 4,435 | 0.34 | 1.08 | 3,901 | 0.30 |
| 12 | 17,372 | 0.72 | 4,368 | 0.25 | 1.24 | 3,714 | 0.21 |
| 15 | 21,715 | 0.77 | 4,304 | 0.20 | 1.38 | 3,545 | 0.16 |
| 18 | 26,058 | 0.81 | 4,241 | 0.16 | 1.51 | 3,391 | 0.13 |
| 21 | 30,401 | 0.86 | 4,181 | 0.14 | 1.62 | 3,250 | 0.11 |
| 24 | 34,744 | 0.91 | 4,123 | 0.12 | 1.73 | 3,121 | 0.09 |
| 27 | 39,086 | 0.95 | 4,066 | 0.10 | 1.83 | 3,002 | 0.08 |
| 30 | 43,429 | 0.99 | 4,012 | 0.09 | 1.92 | 2,892 | 0.07 |
| 32 | 46,325 | 1.02 | 3,977 | 0.09 | 1.98 | 2,823 | 0.06 |

*Source: http://webapps.bbe.umn.edu/fans/*

air distribution through the grain and greater flexibility for drying and storage uses, but are more expensive than aeration ducts. Therefore, many large diameter storage bins are equipped with in-floor aeration pads or ducts for air distribution. Subfloor ducts work well in large diameter round bins as they do not hamper grain removal. Figure 5-14 shows common subfloor duct arrangements for round bins. Flat storages and piles almost always use on-floor half-round or full-round aeration ducts.

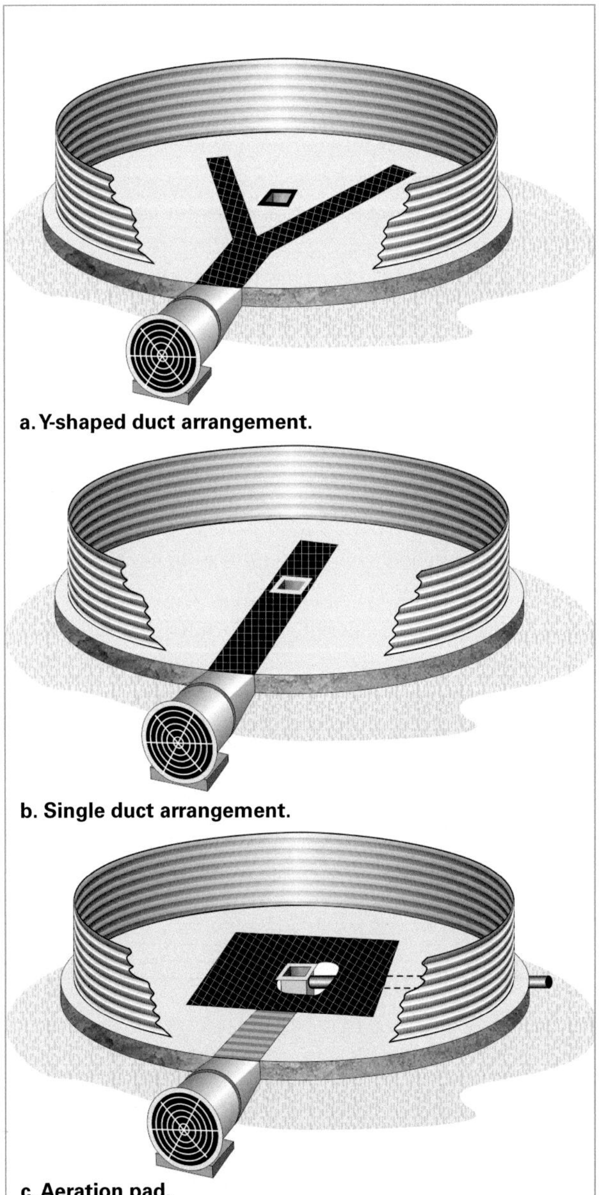

a. Y-shaped duct arrangement.

b. Single duct arrangement.

c. Aeration pad.

**FIGURE 5-14. Common subfloor aeration duct patterns for round bins.**
Duct arrangement may be affected by grain depth. Make the perforated in-floor sections easily removable for cleaning the trench below.

For bins with ducting in the floor, size the duct cross-sectional area so the duct air velocity is less than 1,500 fpm (feet per minute). High duct air velocities increase fan power requirements and can cause non-uniform airflow exiting the duct. Do not exceed 1,500 fpm without proper engineering advice. To achieve 1,500 fpm air velocity, provide 1 $ft^2$ (square foot) of duct cross-sectional area for each 1,500 cfm.

Install above-floor ducts in flat storage structures and outdoor piles for easy removal during grain unloading. Ducts are usually placed directly on the floor and held in place by carefully placing grain along the sides of the ducts so they do not shift when grain is first loaded. Figure 5-15 illustrates how aeration ducts need to be placed above-floor to assure proper airflow distribution in a flat-storage building.

Good judgment is required for duct spacing and arrangement to provide acceptable uniform air distribution. With peaked grain storage, if grain is delivered to the center of the building and allowed to flow to the walls, the grain near the center contains more fines and will be packed more, restricting airflow near the center. In a building up to 40 feet wide, a single, properly sized duct may effectively distribute air, as shown in Figure 5-15a. Buildings between 40 to 60 feet wide with peaked grain require at least two properly sized ducts. Limit duct length to 100 feet per fan. For example, in a 160-foot building use two, 80-foot ducts with a fan on each end of the duct.

For effective air distribution in level piles, duct spacing should be equal to grain depth.

> **For example:** For an 8-foot grain depth, ducts should be 8 feet apart. Place the ducts closest to the wall at a distance equal to half the grain depth from the wall. Figure 5-16 illustrates this concept.

There must also be enough perforated area to allow air to exit ducts without excessive restriction. When using standard perforated metal duct or flooring provide 1 square foot of perforated surface area per 50 cfm. For round ducts, only 75% of duct surface is considered effective. For 3,000 cfm, using round perforated ducts, total duct surface should be: $3{,}000/50 \times 1/0.75 = 80$ $ft^2$ of duct surface area. If the duct has 0.80 $ft^2$ perforated surface per foot of length, a 100-foot length of duct would be required. However, make sure the duct meets other spacing requirements such as length from sides, surface and ends of the storage structure.

Air must have an unrestricted path into and out of the building. Provide at least 1 $ft^2$ of opening in the roof, gable end, or eaves for each 1,000 cfm of airflow. Condensation on roof vents can lead to an accumulation of grain dust in vents over time. This reduces the amount of open area in the

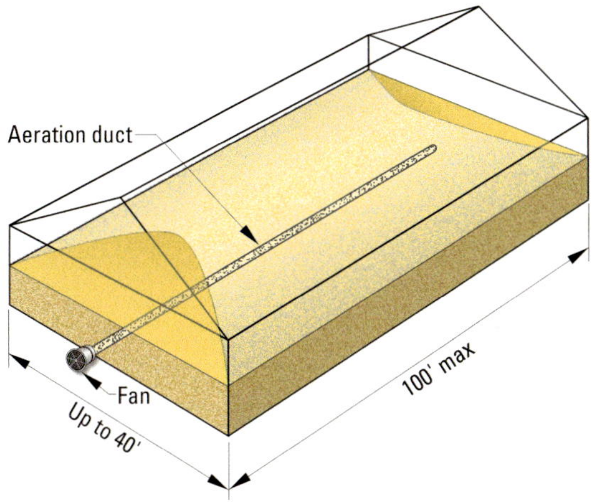


**a. Duct arrangement for building widths up to 40'.**

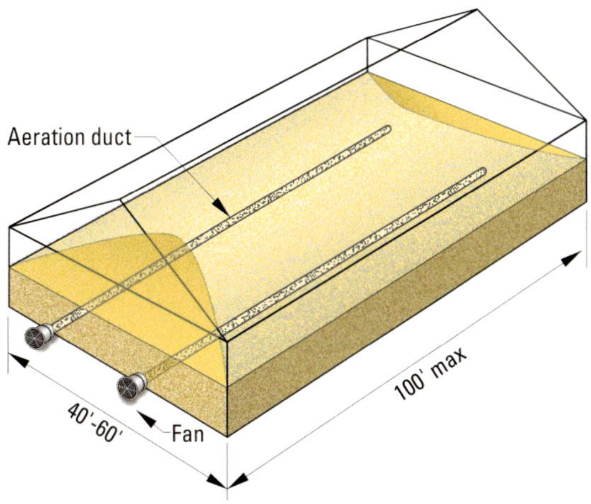


**b. Duct arrangment for building widths between 40' to 60'.**

**FIGURE 5-15 Duct arrangements for building widths up to 40 feet (a) and between 40 and 60 feet (b).** Successful flat storage aeration depends on proper sizing and location of the ducts. Limit duct length to 100 feet per fan. For buildings up to 200 feet long, use two ducts with a fan on each duct and no more than 3 to 5 feet between the closed ends.

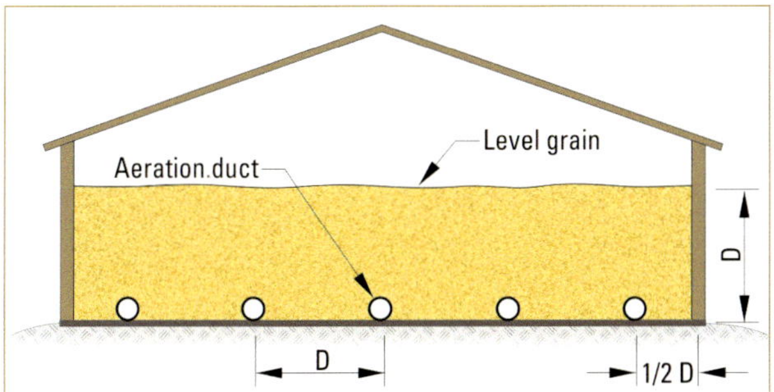


**FIGURE 5-16. Duct arrangement in flat storage with level grain.**

vents, restricts air movement and can interfere with good airflow out of the head space (area between the grain surface and building roof). For this reason, roof vents should be inspected along with other areas of the bin prior to harvest each year and cleaned if necessary to ensure uniform airflow.

In this handbook, general design information for aeration systems has been included. For more specific information about aeration system selection, sizing and layout, see *Dry Grain Aeration Systems Design Handbook*, MWPS-29, or contact your state extension agricultural engineer or a knowledgeable equipment dealer.

## Fines

Broken grain and foreign material, or fines, can create two problems in stored grain, particularly when they accumulate in pockets. First, kernels with severe stress cracks are more susceptible to breakage, and broken kernels are more susceptible to spoilage than unbroken sound kernels. Second, airflow from aeration fans is slowed by pockets of fines, so grain is cooled less unfortunately and more slowly. If cooling is not complete, these pockets often develop into hotspots that result in spoiled grain.

Strive to minimize fines produced from harvesting, drying, and handling, rather than later trying to solve resulting storage problems.

*Consider three techniques for managing fines in stored grain that avoids problems:*

1. **Use a grain spreader** to minimize the concentration of fines in the core of the grain mass. A power spreader needs to be installed and operated properly to avoid a donut-like distribution of fines in the grain mass. The spreader needs to be installed under the bin-fill point to catch the inflow of grain and spread both grain and fines. Make sure the spreader is properly centered under the incoming grain flow; otherwise, the grain surface will not be uniform or close to level.
2. **Clean grain before putting it in a bin** to improve storability, especially if it will be held for 6 months or more. Unfortunately, cleaned grain may not have higher initial market value, but it will allow more uniform air flow and be easier to manage for safe storage. Fines do have value, both as livestock feed and as added weight in market grain. Livestock farmers should clean their feed grain because they can feed the fines and preserve grain quality for longer periods. Unless fines are causing serious storage problems or grade reduction, cash grain farmers may lose money by cleaning grain below the allowable grade level unless they can sell the cleanings to a livestock

farmer or grain elevator. However, partial cleaning may still be worthwhile to avoid problems during long-term storage.

3. **If cleaning grain or using a grain spreader is not practical,** unload a small portion of grain periodically during filling or soon after the bin has been filled to remove the center core where fines and trash accumulate. If left in place, this area of the grain mass will likely develop hotspots during storage unless thoroughly aerated, so should be monitored frequently.

When planning a new grain storage and drying system, plan for a way to include grain cleaning now or in the future because grain markets may change to make on-farm grain cleaning more profitable.

## Grain Peaking

When center filling without a distributor, most dry grain peaks at an angle of 18 to 28 degrees, depending on the type of grain. The volume of grain that can be stored in the peak of a round bin can be estimated with Equation 5-8 or Table A-5 (Spreadsheet S 5-1). Although it is tempting to store those extra bushels, peaked grain is more difficult to uniformly aerate, which can cause heating and subsequent problems.

**EQUATION 5-7.**

$$PC = 0.131 \times D \times D \times D \times SF \times 0.803$$

PC = estimated capacity of the peak of a round bin, bu
0.131 = constant
D = bin diameter, ft
SF = slope factor, Table 5-1
0.803 = conversion factor, $ft^3$ to bu, $bu/ft^3$

**For example:** The estimated capacity of the peak in a 24-foot diameter round bin of corn with an average filling angle of 23 degrees is 610 bushels ($0.131 \times 24 \times 24 \times 24 \times 0.42 \times 0.803$).

Peaking also makes it difficult to check grain from the roof access hole. Entry into a round storage bin with a peaked grain mass is dangerous and should be avoided. If grain is peaked in a round bin during harvest, unload enough grain to remove the peak immediately after harvest. Coring bins just after filling to lower peaks improves aeration management compared to fully peaked bins and makes grain monitoring much easier and safer. Peaked grain has been shown to require two to three time longer to cool than a level grain mass.

## Grain Unloading

For convenient grain bin unloading, use unloading and sweep augers. See Chapter 3 of this handbook for more information about different types of equipment used for unloading grain.

## Insect Protection

Insects require an adequate, readily available food supply in addition to a favorable moisture content and temperature. They are a primary problem in summer-harvested wheat and corn carried into a second storage season. A sound seed coat protects grain from mold and most insects, so sound kernels store much better than do cracked or broken kernels.

Grain fines tend to collect in pockets, blocking airflow and preventing proper cooling. These pockets usually have a higher moisture content than the surrounding grain. This combination increases mold and insect problems.

*Grain damage caused by insects include:*

- Bored holes and the consumption or deterioration of much of the inside of the kernels.
- Injury to the germ.
- Heating and subsequent condensation and grain molding.
- Contamination with insect bodies, insect fragments, excrement, and webbing.

Limit insect damage by controlling grain temperature and moisture content. Insect activity decreases sharply at temperatures below 60°F. Almost all insect activity stops below 50°F. Fumigation is not recommended at grain temperatures below 40°F, because at low temperatures fumigants penetrate the seed coat more slowly. Fumigation should be performed only by a licensed applicator.

*To limit or prevent insect development:*

- Thoroughly clean and spray all storage surfaces with an approved residual insecticide at least two weeks before grain is stored.
- Store grain in weather-tight structures.
- Treat grain with an approved residual protectant to prevent insect infestation in summer-harvested wheat or if corn is to be carried into a second storage season.
- Cool grain with proper aeration.
- Inspect grain at frequent and regular intervals for problem areas.

- Clean fines from aeration ducts and under perforated floors or fumigate with an approved product. Fines provide an excellent breeding place for insects.

## Monitoring Grain

Monitor stored grain weekly during critical fall and spring months when outside air temperatures change rapidly and during summer when temperatures are high. Monitor grain at least every two weeks during winter. Installing at least one temperature cable in the center of each bin is good insurance. Using multiple temperature cables, one in the center and others placed throughout the grain mass provides better knowledge of grain condition.

Consult an equipment supplier for recommended numbers of cables for different sizes of bins. Supplementary temperature monitoring with monitoring for carbon dioxide generation from biological activity (insect-generated hot spots, spoilage onset due to molds) allows for installing fewer temperature cables. This becomes critical as bins with larger and larger diameters require substantial engineered roof structures to carry the load of too many temperature cables.

Monitor every bin at two-week intervals, and check for signs of moisture movement—wet, slimy grain, ice or frost accumulation, crusting, and/or heating. Monitor for odors by turning on the fan and measuring carbon dioxide with a hand-held sensor as well as smelling the air exhausting from the grain, then reseal the fan. A higher $CO_2$ reading from week to week is an early indicator of a storage problem. By the time you smell a foul or musty odor spoilage has advanced beyond onset. Also, check and record grain temperatures every week or two. If using an automatic temperature monitory system set alerts for excessive increases. An increase in temperature also indicates a problem.

## Safety Practices

Routinely review safety precautions and practices with all workers and family members prior to harvest and during storage. Absolutely forbid entry into a bin or gravity unloaded vehicle when grain is flowing or unloading/filling equipment is operating. Ideally, avoid entering grain bins. Suffocation is a major cause of accidental fatality when handling grain. With a flowrate of 1,000 bu/hr from a 6-inch unload auger, a person is knee deep and helpless in 2 to 4 seconds and completely submerged in about 20 seconds after reaching the center of the cone of flowing grain. Before entering a bin, lock out the control circuit on automatic grain unloading equipment, and tag the switch on manual equipment (Lock-Out-Tag-Out) so someone else does not start it by mistake.

Breathing mold spores in stored grain can cause illness and can lead to chronic health problems. Wear a respirator capable of filtering fine dust and mold spores (rated as N-95 or N-99) when working with dusty or moldy grain.

Maintain proper and effective shields and guards on hazardous equipment such as moving belts, chains, pulleys, gears, and shifts.

Chapter 6, Safety, of this handbook provides more detailed information about safety precautions and practices.

CHAPTER 6

# **SAFETY** AND **HEALTH** IN GRAIN DRYING, HANDLING, AND STORAGE

*Mark Purschwitz, Ph.D., University of Kentucky*

The safety and health of people working in and around grain drying, handling, and storage facilities goes well beyond safety training and personal protective equipment. Safety and health (herein called safety) starts with the design and layout of the system and selection of system components. Managers of existing facilities should review their systems for hazards and make changes necessary to improve safety.

The key to any safe workplace, regardless of type, is the control of hazards. Worker training and behavior is important, but identifying, eliminating, minimizing, and/or guarding hazards takes precedence. People make mistakes, regardless of training and knowledge. Virtually everyone has worked when tired, in a hurry, preoccupied, or distracted. Virtually everyone has taken shortcuts or gotten complacent. This is human nature, and the design and operation of grain drying, handling, and storage facilities must take human nature into account. When someone makes a mistake or does something unexpected or improper, the control of hazards is what prevents a serious injury or fatality (herein called injury).

This chapter is directed at owners and managers of existing or future grain drying, handling, and storage facilities. It focuses on planning for a safe facility, and avoiding or controlling hazards, particularly those that can be life-threatening. It also discusses grain bin entry, a leading factor in grain-related fatalities. *It should be noted that all facilities vary and may have hazards or situations requiring management attention that are over and above what is covered here.*

Many factors can be involved in the series of actions or events leading to a serious injury or fatality. Some of these factors may seem to be unrelated to safety. However, injuries often have root causes leading far back in time, and anticipating and eliminating such causes is important. *Again, variability of facilities and situations must be noted; this chapter does not claim to cover every possible factor.*

## OVERALL PLANNING

Anticipating and eliminating as many hazards as possible in the planning stage is the easiest path to a safe facility. Facilitating safety of anticipated hazardous activities such as grain bin entry is easiest in the planning stage. While location of the facility itself has safety implications, the purpose of this section is to inform or remind facility planners of the variety of safety-related issues that should be considered once the overall location has been established.

Within the facility, the location of structures and other system components should minimize hazards and maximize safety. An initial consideration is traffic flow. Properly-designed traffic flow can eliminate the need for backing up trucks and trailers, which creates blind spots and increases the possibility of running over other workers or bystanders. A seemingly harmless act like noticing a low tire at the rear of a trailer and walking closer to observe could become a fatal incident. Driveways designed to facilitate easy turning by large trucks should eliminate maneuvering that can lead to injury or property damage. Chapter 2 of this publication provides more detail about driveway design.

Space for non-routine events, such as the presence of utility or maintenance trucks and personnel, or the removal of snow, should be provided. Emergency vehicles must be able to access all areas of the facility. Proper grading, surfacing, and drainage of driveways is important to prevent muddy areas where vehicles might get stuck and need to be pulled out, a potentially hazardous situation that should be avoided.

Improper location of power lines can result in unnecessary life-threatening hazards, so consultation with the electrical utility is advised. Inadvertent contact with overhead power lines while moving tall equipment (for example, portable augers) or installing or maintaining equipment at heights can cause fatalities. Power lines should be buried (and marked on plans), or if not buried, moved away from potential points of contact. Follow power line clearance rules. Future changes or expansion of the facility should be considered when locating power lines.

Because all equipment will need maintenance and repair, providing safe access to that equipment should be part of the planning process. For example, safe elevated platforms, large enough to work on, and walkways should be included to facilitate these operations, as well as safe stairways or ladders. The need to climb over things to gain access should be eliminated if possible. Safe methods should be provided for climbing onto trucks and wagons if needed. Equipment suppliers should be consulted about equipment and structures needed for safe access.

In addition to component location and safe access, ergonomics and human factors should be considered. Stairways are easier on the body than ladders. Equipment that requires heavy lifting or high forces to operate or move should be avoided if at all possible. Controls and panels should be well-labeled, easily accessible, and designed for intuitive operation. Outdoor areas should have adequate lighting for safe operation during hours of low light; details on lighting levels can be found in the reference material located at the end of this handbook. Plans for work stations, walking areas, stairways, driveways, and loading or unloading areas should consider potentially slippery conditions. Consultation with knowledgeable equipment suppliers and/or local utilities about these issues is recommended.

It is incumbent on planners to visualize all operations and consider the potential for dangerous situations, especially those unique to the facility. It is also incumbent on planners to be sure that all facilities comply with any applicable codes (for example, building, electrical, fire, safety) and insurance requirements.

## EQUIPMENT QUALITY AND AUTOMATION

The quality of equipment and other system components can have a positive or negative effect on safety. Equipment problems often require workers to engage in hazardous activities such as climbing, unplugging, disassembling, or accessing hazardous areas. Any equipment that requires more frequent or difficult adjustment or maintenance may result in these activities being postponed or avoided, leading to bigger problems later and potential injuries. More downtime means more stress and an increased likelihood of hurrying, taking shortcuts, and making mistakes that lead to injuries. Additionally, all electrical and mechanical equipment must be designed for use in dusty conditions, both for durability and to avoid sparks leading to combustion.

Automation (discussed in Chapter 7 of this handbook) can have a positive effect on safety. Using sensors or indicators that minimize or eliminate the need to climb structures or enter bins will greatly reduce the risk of injury. Grain level indicators, sensors that monitor grain temperature and moisture, and sensors that monitor equipment for slowdowns, malfunctions or other failures are all examples of how automation can positively affect safety. As technology improves, more and better remote monitoring of all aspects of the facility should be considered for safety purposes as well as for more convenient operation.

Automation can also bring additional hazards. The benefits of machines that operate automatically on sensor input or timers can be negated if people are not fully warned about sudden or automatic startup. These features make warnings and proper lock-out/tag-out procedures crucial prior to working on or near such equipment. These are discussed later in this chapter.

## PLANNING FOR FALL PREVENTION AND PROTECTION

Falls are an obvious hazard at grain handling and storage facilities. The risk of falls can be minimized though proper planning, installation of proper equipment, and the use of proper procedures. This section focuses on permanently-installed equipment to minimize the risk of falling. This equipment should meet OSHA standards or ANSI (American National Standards Institute) standards, which are newer and more rigorous; both are referenced at the end of this chapter. In addition, OSHA standards and equipment suppliers should be consulted about proper personal protective equipment (PPE) and individual fall arrest procedures required for working at heights.

Installing stairs instead of ladders on bins, bucket elevators, and other tall structures can reduce the risk of falls. Climbing ladders is more tiring and hazardous compared to climbing stairs, especially in wet or icy conditions or when bundled up in cold weather. Stairs must be properly connected to the structure, have proper handrails and no-slip treads, and have a well-guarded platform at the top. Easier access to bin hatches increases the likelihood that grain will be properly checked and managed. As discussed later in this chapter, poor grain conditions are a key factor leading to bin entry and risk of entrapment.

If ladders are used, rungs, side rails, attachment brackets, and fasteners must be kept in good condition and monitored for corrosion. Broken or badly corroded parts must be replaced, and the ladder kept firmly attached to the structure. Cages around ladders, like the example in Figure 6-1, are recommended, along with landing platforms for taller ladders to enable rest during ascent or descent. Also necessary are adequate railings and handholds to facilitate transfer from ladder to platform or walkway.

Once a worker is at elevation, safe platforms and walkways are crucial. Workers are focused on their tasks and not fall prevention. Platforms should be sized for work activities. Bucket elevators supported by towers rather than guy wires not only facilitate the inclusion of steps, but also provide landing platforms that can be used as safe entrance points to catwalks. Proper guardrails, intermediate rails, toeboards, and walking surfaces should always be provided on stairs, platforms, and catwalks. Guardrails meeting OSHA standards are considered fall protection systems, and if inadequate guardrails are used, personal fall arrest equipment would be required, where workers wear harnesses and lanyards connected to anchors of proper design and strength as displayed in Figure 6-2. OSHA standards require fall arrest anchors to withstand 5,000 pounds.

Note that OSHA standards require fall protection at heights above 4 feet, and proper stair railings for flights

**FIGURE 6-1. An example of a cage enclosure on a bin ladder.**

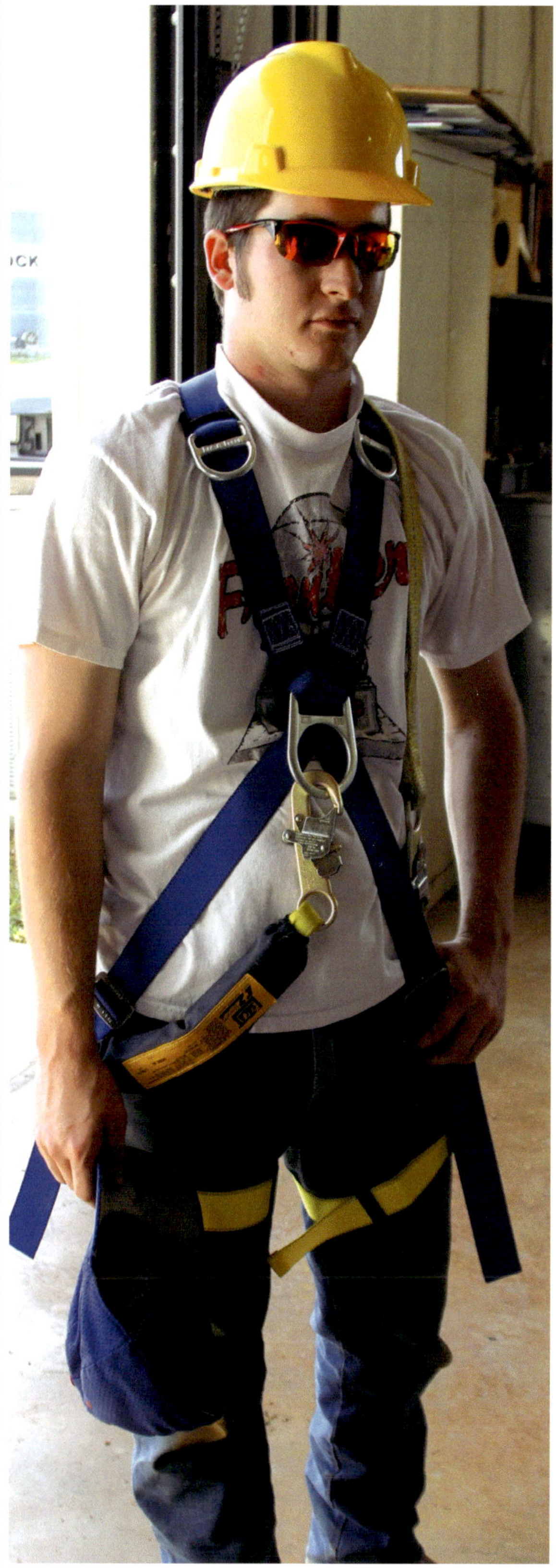

**FIGURE 6-2. Full body harness with fall protection lanyard temporarily attached to front D-ring.**
*Photo courtesy of Grain Entrapment Prevention Initiative.*

of steps with four or more risers. OSHA also requires fall protection when working above hazards like machinery regardless of height. It is important to consult OSHA and/or ANSI standards about fall protection. People can fall from steps, ladders, platforms, and walkways; these standards explain when fall protection and/or personal fall arrest systems must be provided, and describe the acceptable means to do so. Reputable equipment suppliers will have details on steps, ladders, platforms, and walkways designed and constructed to meet the standards. Fall protection must be an integral part of the planning process.

Falls do not occur only from elevation, but also can occur on the ground. This means that any ground-level hatches, access ports, or other openings must be kept closed, or marked and guarded when open, to keep people from inadvertently falling into them. People working at ground level are focused on their own tasks, not open hatches. It also means that broken concrete, protrusions in walking areas, or other tripping hazards must be removed, and proper walkways must be provided. Slips and trips on spilled grain, tools, or general clutter can be avoided by good housekeeping. The need to climb over obstacles should be minimized or eliminated.

## GRAIN CONDITION AS A SAFETY ISSUE

Grain condition is usually considered an economic issue. In fact, good grain condition is crucial to the safe operation of a grain, drying, handling, and storage facility, and maintaining grain condition is actually one of the most important of all safety practices, since two of the three entrapment hazards discussed later in this chapter can be traced back to grain in poor condition. Proper planning or modification of a facility includes the equipment needed to maintain good grain conditions.

The crusting and clumping caused by grain in poor condition leads to climbing ladders or stairs, working on elevated platforms, and entering grain bins. Grain bins are confined spaces, and entry is one of the most hazardous tasks in a grain facility; it should be minimized or avoided if possible. Grain bin entry requires numerous procedures that are beyond the scope of this chapter, and are found in other publications, such as the OSHA standards. Proper handling of fines also has safety implications. If uncleaned grain is loaded into a bin without a grain distributor, fines will tend to accumulate in the core of the grain mass, which interfere with airflow, have higher moisture content, and lead to clumping or crusting; therefore, cleaning grain or dealing with fines can be considered a safety practice.

Grain stored at high enough moisture levels can heat up and lead to a fire. Pockets of fines with higher moisture content can do the same thing. Burning or smoldering grain can create hollow pockets inside the bin, which can in turn put employees and emergency responders at risk of pocket collapse and engulfment when they enter to investigate.

Plans for storage should include adequate volume to avoid grain peaking, i.e., storing grain in the peak of a round bin. Because the peak is more difficult to aerate, it can lead to grain condition problems, which in turn lead to the hazards associated with clumping and crusting. If it is absolutely necessary to enter a bin, peaking makes it more difficult and dangerous, interfering with the safety procedures required (discussed later in this chapter).

## GRAIN CONDITION AND RECLAIM SYSTEMS

Planning the reclaim system is an important aspect of safe grain storage. With grain bins, potential grain condition issues should be considered when selecting reclaim equipment and deciding on the configuration of the foundation. The proper reclaim system can help minimize the need to enter a grain bin. The proper foundation, in combination with the right equipment, can improve overall worker safety.

Zero-entry grain sweeps improve worker safety by keeping workers away from operating augers and related equipment and satisfy the OSHA mandate that workers can enter bins only with the equipment shut off and locked out. In addition, the ability of the overall reclaim system to handle clumps and other grain-condition-related problems without worker entry is another factor to consider when selecting equipment. Equipment rugged enough to operate in poor condition grain without breaking down or requiring entry to unplug is important. The size and number of discharge holes, the grates that cover them, and the type of gates used to operate them, should all be considered in the planning process.

Foundations can be designed with tunnels beneath the floor large enough to accommodate equipment alone or safely accommodate workers as well. The latter applies mostly for larger diameter commercial bins but has the advantage of allowing easier maintenance, making cleanout easier, and, what is more important, providing access to equipment from beneath. This allows dealing with blocked discharge holes, gates, or other problems related to grain clumping without entering the bin. An external opening to the sump can also be provided. A tunnel only large enough to accommodate equipment may be considered a confined space, whereas tunnels designed to easily accommodate workers may not be. Planning for the size of equipment is important to allow room for worker access. Note that codes and standards may be involved in such designs. An engineer should be consulted when planning foundations with tunnels.

## ENTRAPMENT HAZARDS OF GRAIN

Unloading grain with a person inside the bin is the leading cause of grain entrapment fatalities. It can take just seconds for a person to be pulled down into the grain, illustrated in Figure 6-3, where he/she cannot self-extricate, and just a few seconds more to become completely submerged, leading to suffocation. Bins must not be entered while unloading equipment is operating. Walking down the grain during unloading is unnecessary and must be forbidden.

If grain has crusted and bridged, the bridge should be broken up from outside the bin if possible, such as with a long pole, rather than entering and risking a bridge collapse and entrapment, depicted in Figure 6-4. If it is necessary to enter the bin to break up a crust, it must be done only while the unloading equipment is shut off and locked out and the entrant is wearing a body harness connected to a properly installed and controlled lifeline, as discussed in the next section. To prevent flowing grain entrapment, breaking up buried clumps must be done only when the unloading equipment has been turned off and locked out. Again, the entrant must be connected to a proper lifeline.

Entry to break up crusted grain overhead, such as old grain that has become attached to the wall, must be forbidden. Figure 6-5 depicts how grain dislodges in such situations can collapse and become an avalanche that

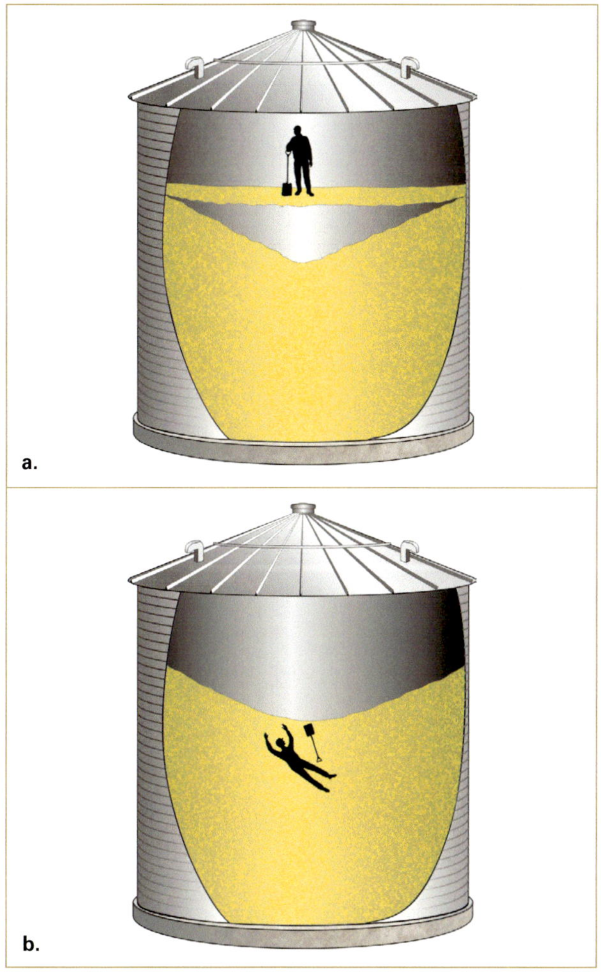


**FIGURE 6-4. A second way of becoming entrapped in grain involves entering to break up a bridged crust (a), and the bridge collapses (b).**

**FIGURE 6-3. A worker can quickly become entrapped in flowing grain when unloading equipment is operating.**

**FIGURE 6-5. A third way of becoming entrapped in grain occurs when grain stuck to the bin wall releases, and the avalanche of grain strikes and/or covers and entraps the person below.**

entraps the person or persons beneath. Even if the grain does not totally collapse, large, heavy clumps may fall and strike people below, seriously injuring or killing them. Breaking up grain attached to the sides of bins should be done only from above. Better yet, grain should never be allowed to get in such poor condition and should not be left in a bin for long periods of time during which this kind of crusting can occur.

It should be noted that an option given in the OSHA grain handling standard is use of a boatswain's chair, shown in Figure 6-6. However, procedures and equipment needed for use in grain bins are not spelled out in the standard and have not been developed. Future development, in conjunction with grain bin manufacturers, is possible, but even then, obtaining the proper equipment and training would be critical. Until proper procedures, equipment, and training are available, the boatswain's chair is not recommended and is mentioned here only as a potential future option.

**FIGURE 6-6. Boatswain's chairs require use of a personal fall arrest system in conjunction with other suspension equipment.**

## PLANNING FOR POSSIBLE GRAIN BIN ENTRY

**Grain bins should never be entered unless absolutely necessary;** the goal of any grain facility manager should be zero entries. It is well known that entering a grain bin or other storage facility can lead to grain entrapment injury and fatality. As discussed previously, sensors and other automation can reduce the need to enter, as can proper reclaim systems.

However, in anticipation of times when entry may be necessary, proper planning prior to construction can greatly improve the safety for such procedures. It should be noted that complete confined space entry procedures, including preplanning, permitting, testing the atmosphere, respiratory protection, and emergency procedures, are not the subject of this chapter and are covered by other sources and publications.

### Lock-out/Tag-out

Prior to any entry, it is imperative that all loading and unloading equipment be locked out and no grain be flowing. All internal equipment, such as stirring augers or sweep augers, must be locked out to prevent entanglement. Equipment that starts automatically also must be turned off and locked out to prevent startup. All equipment involved in moving grain, such as grain legs and cross conveyors, must be locked out. In the planning stage, it is important to provide and locate electrical boxes and/or controls in easily accessible places to facilitate lock-out/tag-out procedures. Inconvenient locations tempt workers to skip the procedure.

Proper lock-out/tag-out procedures require each person working on hazardous equipment or working in a hazardous area (such as a bin) to have his/her own padlock with a single key under their control only. Workers needing to lock out multiple controls will need multiple locks, each with a key under their control, or keyed together in a set. Nobody can use a locked-out switch until every single worker is out of harm's way and has personally removed his/her individual lock. A tag, like the one in Figure 6-7 identifies the owner of that lock.

Trying to protect multiple workers with a single lock is dangerous and not acceptable, because it is possible to remove the lock while one worker is still exposed to harm. Tag-out procedures alone (tagging equipment switches but not locking them) are unacceptable, because tags can be removed by someone other than the person inside the bin. Lock out physically prevents someone from energizing the equipment.

Planning for proper lock-out/tag-out procedures should include making the locks and tags readily available to workers. Proper storage areas for this and other personal protective equipment should be provided in a convenient location. Lock-out/tag-out is a fundamental safety practice that should be included in and facilitated by good planning.

### Lifelines

A tenet of safe grain bin entry is proper use of a lifeline designed to prevent grain entrapment. A rope wrapped around the waist or shoulders is inadequate and unsafe, and should not be used. Personal fall arrest systems, designed to prevent injury from falls, are not designed to prevent entrapment, and should not be confused with grain bin lifelines.

A lifeline for grain bin entry requires special equipment and strong anchor points. It is much easier to plan and build this into a bin than to add it later, although retrofitting existing systems should still be considered. Lifelines require two anchor points, one overhead and another at the sidewall. Locations of anchor points and the anchors themselves must be able to withstand the forces exerted by lifelines. Some bins are not strong enough to withstand roof loads necessary for anchor points. Installing anchor points should be done using proven methods and equipment and in consultation with the bin manufacturer, an engineer, or other qualified person.

The forces on a person becoming entrapped in grain can be very large. Tests with mannequins submerged in flowing grain produced downward forces averaging 800 pounds when the mannequin's head was just below the surface. Forces were less for mannequins partially submerged, and greater for mannequins more deeply submerged, nearly 1,800 pounds. Neither ladders—including rungs, side rails, and brackets—nor railings on stairways and platforms, are strong enough to serve as anchor points and should never be used. Engineered anchor points must withstand at least twice the anticipated load, and typically are stronger than that, depending on type and location. A future American Society of Agricultural and Biological Engineers (ASABE) standard under development for new grain bins will address anchor strength and other entry-related safety features.

One example of an overhead anchor point, on a bin designed to withstand the forces involved in a person being trapped, is shown in Figure 6-8. This system uses knot-passing pulleys, as shown. The lifeline itself would require another anchor point on the bin sidewall. The key is to have a system that can withstand high forces and still enable a worker to move around without becoming entrapped. As indicated earlier, this is a dedicated system to prevent entrapment and is not a fall arrest system.

Workers entering a bin at or above grain level must be wearing a full body harness attached to the lifeline. The lifeline itself includes certified ropes, D-rings, belaying systems, and other components that require knowledgeable selection, training, and use. An observer must be stationed outside the bin, on a platform with fall protection; workers

**FIGURE 6-7. Example of a properly locked-out electrical box, including padlock and tag.**
*Photo courtesy of Mark Purschwitz.*

**Figure 6-8. Example of a proper overhead anchorage point on a grain bin designed to withstand the forces involved when a person is entrapped.**

should never enter a bin alone. The job of the observer is to continuously watch the person inside, stay in verbal communication, and feed out or take in slack on the lifeline, which is crucial to prevent potential entrapment. The observer can also get help if needed, although it is strongly recommended that a third person be present on the ground and available to get help.

Details on the equipment, procedures, and training needed to install and use a grain bin lifeline are beyond the scope of this chapter and can be found elsewhere, including the supplied references. However, planning for bin entry is an important part of the overall process of planning a safe grain drying, handling, and storage facility.

## PLANNING FOR MACHINERY SAFETY

Grain drying, handling, and storage facilities use many types of machinery with hazards that must be controlled. Facility planning should include the selection of machines that meet OSHA and/or ASABE safety standards; the equipment supplier should be able to verify this. Comparisons of machines may show that some manufacturers have eliminated hazards by design. Shields on machines should be designed with maintenance in mind. Shields and other safety features that are difficult to use, remove and replace invite workers to save frustration and leave them off.

Belts and chains and the corresponding pulleys and sprockets must be shielded. Auger flighting must be covered or protected by guards. Rotating shafts must be guarded; tractor PTO–driven equipment must have shields on the tractor, driveline, and driven machine. All shields must be in good condition, function properly, and must be put back in place if removed for maintenance. Planning for machinery safety includes not only following instructions on warning labels, but replacing labels that have become damaged or otherwise illegible. Figure 6-9 displays a good example of a warning label. Plan to provide hearing protection to workers near noisy machines or processes. OSHA standards provide information on noise and hearing protection.

Vacuum grain handling systems can be dangerous if used improperly, and workers should be trained in proper operation. The primary hazard in using a vacuum system involves workers removing grain from beneath themselves or another person, resulting in an entrapment. Removing grain beneath a person is like creating quicksand, and it can happen so quickly that a person is soon engulfed. It is a known source of fatalities, particularly in flat storage structures.

When tractors are used as stationary power sources for portable machines, precautions must be taken to assure they do not move. If the driven machine has a hitch, it should be

**FIGURE 6-9. Example of a warning label for machinery; should be replaced if damaged or illegible.**
*Courtesy of Sukup Manufacturing Company.*

connected to the tractor to prevent the tractor and machine from moving apart. A PTO driveline that separates while in operation is extremely dangerous. Wheels of the driven machine should be blocked, and unless the tractor's transmission has a "park" position, it should also be blocked. The tractor and machine should be positioned such that people can walk around them and not have to step over the PTO driveline. Safe refueling procedures must be followed.

If a skid-steer loader is used to move grain in a pile or flat storage building, the appropriate safety procedures must be followed. Safety procedures for tractors and skid-steer loaders are beyond the scope of this document, but will be found in operator's manuals.

A high-temperature grain dryer is a machine with unique safety hazards. Fires must be prevented, hearing protection must be provided to workers in the area, and drive mechanisms must be properly guarded or shielded.

Fire hazards may involve several factors. One is the crop being dried. Constant supervision is necessary with high-temperature drying of sunflowers and grain sorghum. Another factor is trash accumulating in the drying column that becomes dry and combustible. Assure that grain flows in all parts of the column and periodically clean inside the column to remove any trash accumulation. Trash can also be pulled into the dryer fan and burner. Trash around the dryer should be removed daily. A third factor is fuel handling and storage. Any LP gas or other fuels must be stored safely away from the dryer and where tanks will not be hit by traffic. Underground lines should be marked on plans.

## PLANNING FOR MAINTENANCE

All machines and equipment require some degree of maintenance, service, and repair (herein called maintenance). Because injuries often occur during maintenance, planning should include facilitating safety during those procedures. Worker training is important, but facility design can improve safety during maintenance.

Safe access to maintenance points must be provided, consistent with earlier discussions on fall prevention and grain bin entry. The need to crawl over equipment or pass through unsafe areas should be minimized. Workers at heights must be protected against falls, either by proper railings or by wearing a personal fall arrest system connected to a proper anchor. The area below should be blocked off to prevent injury from falling objects, and workers below should wear hardhats. An observer on the ground should be present to call for help if needed.

As discussed earlier in this chapter, all machines undergoing maintenance must be locked out by the worker(s) doing the job. This includes any related equipment, or physically close equipment, that could create a hazard if operated. Equipment that starts automatically is particularly hazardous and must be locked out. Other equipment that may create hazards to workers, such as grain loading or unloading equipment that may be activated during maintenance activities inside a bin, must also be locked out.

Large portions of the electrical system may have to be deactivated and grounded during work on breaker boxes, control panels, or other parts of the electrical system where unexpected current flow may occur. Any other unexpected release of energy, such as pressurized gas, must be prevented. Lock-out equipment for valves and other controls is available from safety equipment suppliers.

Plan on making all necessary tools and PPE available. Using poor-quality tools, makeshift tools, or worn or damaged tools increases the difficulty of the job and increases the risk of injury. Any special equipment or tools needed for maintenance procedures, including for safe access and working at elevation, should be considered in the planning process. Proper PPE must be provided and required to be used. Training on proper tool usage and shop procedures should be provided; this information is outside the scope of this chapter, but it is readily available elsewhere.

Any welding or other hot work should only be done by trained workers using proper equipment and following proper procedures. Fire hazards must be removed, especially grain dust and trash. The path of sparks, for hot work done at heights, must be considered. Fire extinguishers must be available.

The aforementioned risks and safety procedures may also apply to installing or removing structures and other equipment. Any interaction between people and machines must be done with proper planning and preventive measures.

Plan on keeping good records. Any inspections, repairs, equipment changes, unusual problems or findings, or anything connected to maintenance should be documented. This will be helpful later if the same problem occurs or if there are warranty issues. Availability of proper tools and PPE should be documented. Safety procedures related to maintenance, such as lock-out/tag-out and confined-space entry, should be documented. These records are all part of institutional memory, which is important to be recorded and not just kept in people's heads. Personnel may change; memories may fade or get confused, and what seemed obvious might be unclear later. In addition, good records may be crucial if liability issues arise, particularly if there is any question about how equipment was maintained or if proper and safe procedures were followed.

## PLANNING FOR DUST AND CHEMICAL HAZARDS

Grain dust is an inevitable part of grain handling and storage. It poses a respiratory hazard in high concentrations, and even in low concentrations it can cause problems for people with existing respiratory problems. Studies have shown that exposure to grain dust which can contain mold spores can cause "grain fever", wheezing, chest tightness, cough, eye, nasal and skin irritation. People with serious allergies to molds may have more severe reactions. Proper PPE should be used to minimize exposure to grain dust and mold spores. Facility plans should include readily accessible clean storage locations for respiratory protection (approved dust respirators) in quantities adequate for all workers exposed to grain dust, plus visitors. Workers who experience respiratory problems or flu-like symptoms after exposure to grain dust should see a physician and explain the exposure.

Grain dust is also an explosion hazard. Explosions in commercial grain elevators are well documented and typically involve poor dust control in combination with a spark or other heat source. Keeping dust from accumulating or removing accumulated dust, particularly in confined areas, is important. This also reduces the respiratory hazard of accumulated dust becoming airborne. Proper maintenance of machinery to eliminate problems from overheating or sparking is important. These procedures are well covered in other publications.

Use of chemicals, such as fumigants, can be dangerous. If fumigation is anticipated on a regular basis, it should be included in the planning process. Besides requiring an applicator license and a written plan with proper procedures, fumigation requires proper PPE and the ability to seal a bin, all of which should be anticipated. The label, depicted in Figure 6-10, is a legal document and must be followed. Only workers trained and licensed in the use of fumigants should be allowed to work with them. Professional application is an option if training and certification are not available or sought ahead of time.

RESTRICTED USE PESTICIDE
DUE TO HIGH ACUTE INHALATION TOXICITY OF PHOSPHINE GAS
FOR RETAIL SALE TO DEALERS AND CERTIFIED APPLICATORS ONLY. FOR USE BY CERTIFIED APPLICATORS OR PERSONS UNDER THEIR DIRECT SUPERVISION, AND ONLY FOR THOSE USES COVERED BY THE CERTIFIED APPLICATOR'S CERTIFICATION. REFER TO THE DIRECTIONS IN THIS APPLICATOR'S MANUAL FOR REQUIREMENTS OF THE PHYSICAL PRESENCE OF A CERTIFIED APPLICATOR.

THE COMPLETE LABEL FOR THIS PRODUCT CONSISTS OF THE CONTAINER LABEL AND THE APPLICATOR'S MANUAL WHICH MUST ACCOMPANY THE PRODUCT. READ AND UNDERSTAND THE ENTIRE CONTAINER LABEL AND APPLICATOR'S MANUAL.

A FUMIGATION MANAGEMENT PLAN MUST BE WRITTEN FOR ALL FUMIGATIONS PRIOR TO ACTUAL TREATMENT.

CONSULT WITH YOUR STATE LEAD PESTICIDE REGULATORY AGENCY TO DETERMINE REGULATORY STATUS, REQUIREMENTS, AND RESTRICTIONS FOR FUMIGATION USE IN THAT STATE. CALL 540-234-9281 / 1-800-330-2525 IF YOU HAVE ANY QUESTIONS OR DO NOT UNDERSTAND ANY PART OF THIS LABELING.

APPLICATOR'S MANUAL
FOR

**Fumitoxin**

TABLETS AND PELLETS

FOR USE AGAINST INSECTS WHICH INFEST STORED COMMODITIES AND CONTROL OF BURROWING PESTS

**Active Ingredient: Aluminum Phosphide** ............ **55.0%**
**Inert Ingredients:** ............................................ **45.0%**
**Total** .......................................................... **100.0%**

KEEP OUT OF REACH OF CHILDREN
DANGER - POISON - PELIGRO

FOR BURROWING RODENT APPLICATIONS: THE USE OF THIS PRODUCT IS STRICTLY PROHIBITED WITHIN 100 FEET OF ANY BUILDING WHERE HUMANS AND/OR DOMESTIC ANIMALS DO OR MAY RESIDE ON SINGLE OR MULTI-FAMILY RESIDENTIAL PROPERTIES AND NURSING HOMES, SCHOOLS (EXCEPT ATHLETIC FIELDS), DAYCARE FACILITIES AND HOSPITALS.

**PRECAUCION AL USUARIO**: Si usted no puede leer ingles, no use este producto hasta que el marbete le haya sido completamente explicado.
**(TO THE USER:** If you cannot read English, do not use this product until the label has been fully explained to you.)

Manufactured for:

D & D HOLDINGS, INC.
P.O. Box 116
153 Triangle Drive
Weyers Cave, VA 24486 USA
Telephone: (540)234-9281/1-800-330-2525
Fax: (540)234-8225
Internet: www.degeschamerica.com
E-mail: degesch@degeschamerica.com
EPA Est. Nos. 073926-CHN-001, 073927-CHN-001
5857-NC-001 or 40285-VA-01
EPA Reg. Nos. 72959-1 and 72959-2

Form 36866 (11/2015)

**FIGURE 6-10. A page from a manual for applicators of aluminum phosphide fumigants.**

# PLANNING FOR SAFETY AROUND THE FACILITY

As mentioned in the initial section on overall planning, proper layout of driveways and other traffic areas is a safety factor. It goes without saying that nobody, including workers, visitors, or bystanders, should be struck by traffic around grain facilities. Generous use of bollards embedded in concrete, as seen in Figure 6-11, can protect pedestrian areas as well as fragile equipment from traffic. Planning should consider increased traffic at harvest. It may be necessary to install fences or other barriers to keep people away from hazardous areas. Adequate exterior lighting is necessary for safe operations at night.

Planning should also include proper signage. Figure 6-12 shows a sample of the type of danger and warning signs that should be posted at key points around the facility. These signs should warn people about grain bin entry, climbing stairs or ladders, machinery hazards, control panels and switch boxes, and any other serious hazards or hazardous areas. Signs should warn unauthorized people to stay away. Besides helping to remind workers and expected visitors of hazards, warning signs can help protect the owner/operator against legal liability. Signs should be bilingual as necessary and include additional languages if appropriate. Signs should be easily readable and durable and should be replaced when severely weathered or damaged.

If guy wires are used for bucket elevators or other tall structures, they must be clearly marked with brightly colored sheathing to attract attention and help prevent people or vehicles from hitting them. Figure 6-13 illustrates

**FIGURE 6-11. Protective bollards.**
*Photo courtesty of Mark Purschwitz.*

**FIGURE 6-12. Example of a danger label for grain bins; should be replaced if damaged or illegible.**
*Courtesy of Sukup Manufacturing Company.*

**FIGURE 6-13. Brightly colored covers for guy wires.**
*Photo courtesty of Mark Purschwitz.*

this practice. A worker performing tasks away from the elevator, such as mowing the grass, may not be thinking about guy wires. Operators of tall equipment need to notice guy wires before passing beneath them.

Plan on keeping children away from grain facilities. Many components of grain drying, handling, and storage systems are attractive to curious children. Ladders and steps are attractive, and safety barriers may be necessary. Other attractive hazards include equipment in operation and trucks or wagons. Also, children are at increased risk from traffic because they may not understand the necessity of staying out of the way. Depending on the location of the facility, it may be necessary to fence it off to keep children out, or perhaps install fencing around the house and yard to keep children in.

Children should not be allowed to ride on or play in gravity wagons or grain trucks, whether full or empty. Once unloading begins, a child can be pulled under the flowing grain, just as in a grain bin. Children in an empty truck or wagon can be quickly covered during the loading process, especially when newer equipment with high flow rates is being used.

## PLANNING FOR EMERGENCIES

An emergency plan should be created. Workers and others, such as family members, should understand the plan and be trained on responsibilities and actions. This should be decided before an emergency occurs, not during it. One important component of an emergency plan is a list of emergency contacts. People should be trained on how to contact emergency responders. Emergency contact phone numbers and clear directions to the facility should be posted in prominent places. In a panic situation, people may not think clearly, or it may be a visitor who must make the call. At least one person should be trained in first aid. Emergency-related equipment, such as first aid kits or properly-charged fire extinguishers, should be readily available. Creating emergency plans is beyond the scope of this document, but information is readily available.

Emergency responders need training to properly respond to grain facility emergencies. Facility owners or managers should contact local responders to learn if such training has been done, or to strongly encourage it. Such training includes everything from understanding the hazards of stored grain and its associated equipment to high-angle rescue, grain bin entry, and lock-out procedures. Grain facility owners/managers can support emergency personnel by giving tours and explaining their systems. They can also support training by making the facility available for training exercises when feasible, and even providing old bins for training. Financial support for specialized equipment like the grain rescue tubes shown in Figure 6-14, which have saved lives, would be welcomed.

## OSHA STANDARDS AND THE GENERAL DUTY CLAUSE

Federal OSHA standards (regulations) applying to farms are found in Part 1928 of the OSHA standards, Occupational Safety and Health Standards for Agriculture. These are readily available online and are outside the scope of this document, but should be read by all farmers. Standards applying to commercial grain operations are found in Part 1910.272. However, the safety practices covered in 1910.272 are also valuable to on-farm facilities. Following OSHA standards is one way to improve the safety of any farming operation.

Federal OSHA standards also include recording and reporting work-related injuries, as explained in Part 1904. These standards apply to all employers, but certain exemptions exist, as explained in 1904.1.

OSHA regulations may be enforced by the federal government or by a state government, depending on the state. Over half of the states have state OSHA agencies (which may or may not have "OSHA" in the name) that operate under state plans approved by federal OSHA. State OSHA plans may have additional or more restrictive standards, and/or more widespread enforcement. For example, while federal OSHA enforces standards only on farms with 11 or more employees at any one time, some states enforce OSHA standards on farms with one or more employees. The federal limitation of 11 or more employees involves a Congressional appropriations rider that could be eliminated at any time; plus, it does not apply to states using their own funds for enforcement on small-farm operations. Regardless of whether grain facilities are commercial or on-farm, owners/managers should contact their state labor department or county Extension agent to find out more about OSHA operations and requirements in their states.

OSHA standards include what is called the General Duty Clause (GDC). The GDC includes a statement that employers "shall furnish to each of his [her] employees employment and a place of employment that are free from recognized hazards that are causing or are likely to cause fatality or serious physical harm to his [her] employees." Citations may be issued for failing to take reasonable steps to prevent or abate such hazards. For agricultural employers, this means that even though a particular hazard is not covered by the Part 1928 agriculture standards, the employer still has the

**FIGURE 6-14. Six examples of coffer dams or rescue tubes for stabilizing and removing grain from around a victim entrapped in a grain bin.**

general duty to protect workers from the hazard. This is especially important if the hazard is covered by the OSHA General Industry Standards, found in Part 1910. This means, for example, that while the exact rules for grain bin entry required at commercial operations may not be required on farms, a similar effort must be made and documented to protect farm employees from the hazards of bin entry. Operators of all grain storage and handling facilities on farms are encouraged to read the OSHA 1910.272 commercial grain handling facilities standard.

The Part 1928 agriculture standards specifically identify several Part 1910 standards that apply to farms. One such standard is the 1910.1200 Hazard Communication Standard. While this standard is primarily aimed at informing workers about hazardous chemicals, explosive materials, including grain dust, are also covered. Besides being potentially explosive, grain dust can cause health problems. A 1987 OSHA letter of interpretation (found on the OSHA website, www.OSHA.gov) confirms the applicability of 1910.1200 to grain dust. As a result, employers must inform workers about the hazards of grain dust. Some cooperatives and commercial grain facilities have Safety Data Sheets (SDS) available for grain dust.

## SAFETY MANAGEMENT

Like anything else of importance on a farm, protecting the safety and health of workers and other people requires good management. It carries a moral responsibility to the people involved as well as legal responsibilities and liabilities. It cannot be ignored or considered trivial.

There are numerous safety and health issues that a good manager must cover. Every facility needs a comprehensive safety program and written plan for identifying and controlling hazards, training workers, and responding to emergencies. Such programs and plans are outside the scope of this chapter, but are available elsewhere. This chapter should not be considered to be an all-inclusive list of hazards for grain drying, handling, and storage. Instead, it is incumbent on owners and managers to be aware of safety standards and inspect facilities for all types of hazards. Periodic follow-up inspections to assure compliance and identify new hazards are an important management function.

Good training requires planning. Workers must be informed about all hazards, the risks of injury or illness, and how to perform operations safely. Managers should follow up to be sure the training has been effective and proper procedures are being followed. Management is responsible for setting, practicing, and enforcing safety rules.

Hiring the right workers or assigning tasks to the right people is an important part of safety management. If family members are involved in the operation, a determination must be made about abilities to do jobs safely. This is especially important with younger workers, who may seem capable of operating certain machines or performing certain tasks but who may not have the judgment or emotional maturity to handle problem situations. Management is responsible for determining which workers can do a job safely regardless of the situation.

Managers are responsible for providing the proper PPE and seeing that it is used. In addition, the PPE must be made readily available and kept in good condition. Something as simple as a portable cabinet or covered plastic tub can be useful in keeping PPE clean and organized. Good quality, clean PPE that is readily available when and where it is needed is much more likely to be used.

Managers should always strive for clear and open communication with workers and should also promote an atmosphere where communication from worker to management or worker to worker is welcomed and considered important. Managers need workers to feel free to point out hazards. Workers need to communicate with each other to avoid situations where one worker inadvertently injures another.

Managers should keep good records. Not only does this involve documenting maintenance-related issues and procedures as described earlier, it involves keeping records of anything related to personnel, including but not limited to hiring, training, rule enforcement, and PPE availability and use. Records are but one aspect of good personnel management. Additionally, records can be very important if any injury or liability issues arise.

People are the most important resource of any operation. Management has a moral obligation to protect the safety and health of all involved. For some operations, investing in safety can have a direct financial return in the form of reduced insurance rates and less downtime. For others, investing in safety is like spending money on insurance; it may not bring a return on investment, but it is a necessary part of business. Either way, management is responsible. Some responsibilities can be delegated, but as the old saying goes, "The buck stops here."

CHAPTER 7

# **AUTOMATION** IN GRAIN DRYING, HANDLING, AND STORAGE

*Carol Jones, Ph.D., Oklahoma State University*

Grain drying, handling and storage have undergone changes due to great advancements in automation and computer technology during the past several decades. And changes will continue rapidly in the years ahead. As far back as the 1970's, significant efforts to improve efficiency and control of grain quality and handling have challenged companies and farmers to move in the direction of automatic controllers, variable frequency drives, and monitors to move grain, monitor and control atmospheres within grain storage facilities, and make grain facilities more productive, safer, and more profitable.

## GRAIN HANDLING

As grain enters a storage facility, the first example of automated tasks is most likely weighing and then sampling. Many commercial facilities use electronic scales and automatic sampling equipment, like that shown in Figure 7-1, to draw samples of grain from truck beds and transfer the samples into equipment that analyzes the grain for quality and moisture content. The grain in the truck is then unloaded at the dump pit, and automatic controls send the grain to the proper bin for storage as designated by the control room in the company's offices or by applications on smartphones or tablets. Empty trucks exit to the outgoing scale where truck weight is recorded into the grain management and accounting system. The truck driver receives a ticket as hardcopy or by e-mail, and then the truck exits the facility.

**FIGURE 7-1. Automatic probe sampler ready to draw a sample from an incoming grain truck.**

## GRAIN DRYERS

Automatic controllers are used in grain dryers to adjust and record temperature, air flow, heating times, and moisture content as the grain passes through the dryer. Moisture content data before and after drying can be retained in a database for inventory control and traceability of the product through the handling system. As the controller monitors both grain and ambient conditions, the grain is more accurately dried to specifications because the controller is dedicated to the purpose. It does not require a manager to guess at the time, temperature, and unload conveyor speed required to dry the grain to the desired moisture content.

## GRAIN STORAGE

One of the key advantages that electronics and automation provide to the management of grain quality is the monitoring of temperature with cables equipped with sensors throughout the stored grain mass. An increase in temperature can mean deterioration of grain or microbial activity in the area around the cable sensors. Detection of this increase in temperature gives managers an early warning sign, and the decision to remedy the problem or move the grain can then be made before the problem grows larger and disaster strikes. Many grain bin entrapments can be attributed to grain going out of condition.

**For example:** Chunks of grain do not always go through the unloading system, and wet grain can stick to bin walls. Crusted grain can cause a bridge to form with a cavity under the top surface during bin unloading. All these conditions require remediation, and many times workers enter bins to fix these problems. When a wall of grain collapses or a bridge breaks, a worker can become entrapped. Too many times these situations end in fatality. (For a more in-depth discussion of safety issues with descriptions and illustrations of how people become entrapped in grain, see Chapter 6 of this handbook).

Using temperature cables to detect grain conditions can help reduce the need for workers to enter bins, thus reducing the opportunities for incidents.

## TEMPERATURE MONITORING

The temperature data from cables can be read manually with a handheld monitor, or the cable data can be read automatically at set intervals by a computer system, and the information can be recorded to a database. Figure 7-2 features a computer screen displaying the results of temperature cable monitoring in one bin, and Figure 7-3 illustrates how wireless transmission capabilities can be installed to simplify maintenance in larger facilities. More advanced controllers can integrate temperature monitoring into fan control systems. When relative humidity and temperature conditions meet preset control levels, the aeration fans turn on and then turn off when the conditions in the grain have met the desired levels or the air outside has become unfavorable (too humid or too dry, too warm or too cool) for proper grain aeration. Some cable systems can gather relative humidity data to help estimate grain moisture content. The software within the controller takes the relative humidity and temperature data and converts it into moisture content information that matches the temperature and relative humidity conditions in the grain. $CO_2$ sensors have also been integrated in some automatic monitoring systems for early detection of microbial activity, insect, or deteriorating grain condition. Another feature available for monitoring the conditions in grain storage are electronic insect detectors. These detectors show promise by counting the number of insects passing through a sample tube, and some detecting systems can estimate what kind of insect is present.

Figure 7-2 shows a screen representation of the temperature cables within one bin. Sixteen cables are located at different places within the bin. Cable 1 has 18

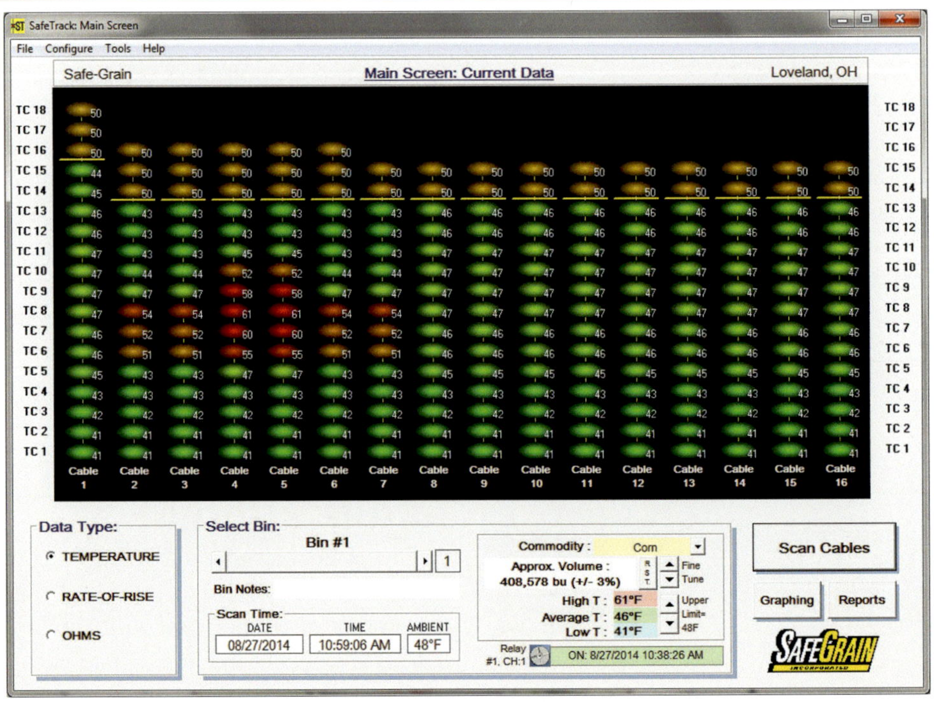


**FIGURE 7-2. Example of temperature cable monitoring in one bin.**

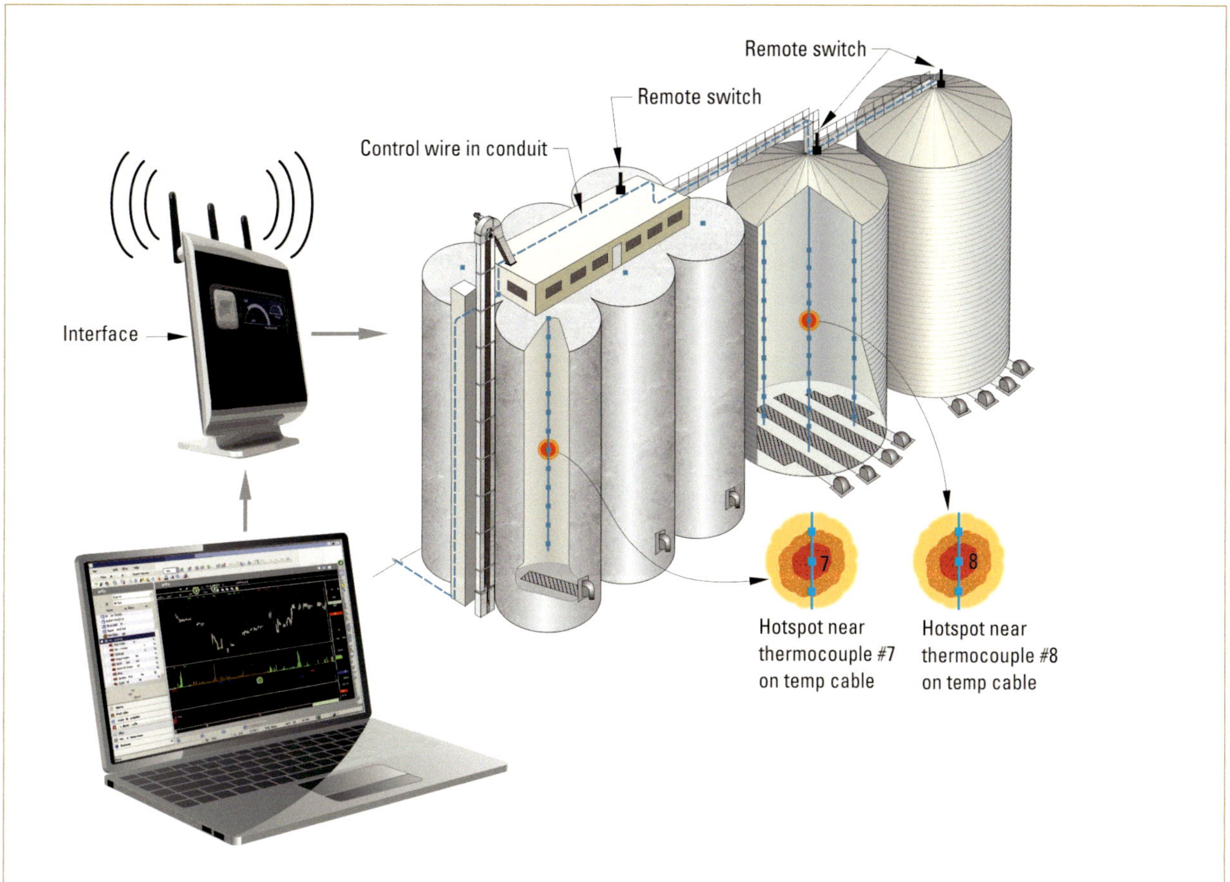


**FIGURE 7-3. Example of bin temperature cables monitored with a wireless system.**

thermocouples; cables 2 through 6 have 16 thermocouples, and the remainder of the cables have 15 thermocouples. Thermocouples are typically spaced at 4 to 6 feet intervals along a cable. A manager could determine three important things from this screen. The grain level is at the height between thermocouple 15 and 16 on cable one and one thermocouple lower on the remainder of the bin. That is where the headspace temperature differs from the grain temperature. The screen also highlights a hotspot or trouble area, shown in Figure 7- 4, where the temperature is higher in the area around cables 4 and 5 at the height of thermocouples 7 and 8 and radiating around that area for a short distance. The remainder of the grain mass in the bin has stabilized at temperatures between 41 and 47°F. The manager can set the alarms at a temperature that would indicate a problem area. This same system can generate reports and graphs of the temperature history by cable. It also could be connected to the aeration fan system to turn the fans on and off at preset temperatures, and if the system is connected to a humidistat system, relative humidity sensors, moisture conditioning can also be considered in the controls for the aeration fan system.

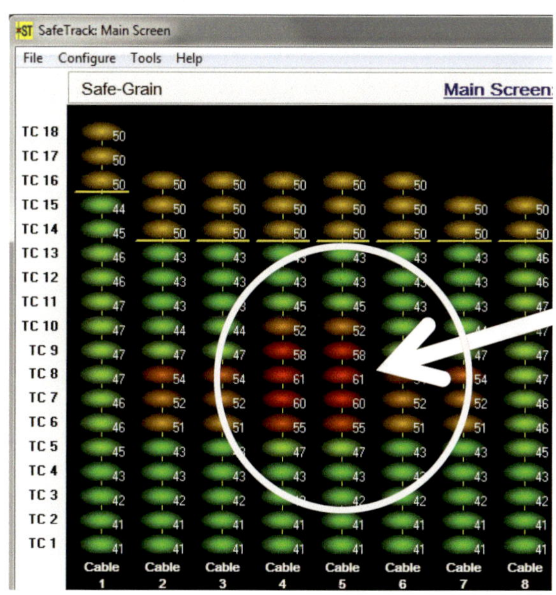


**FIGURE 7-4. Thermocouples on temperature cables indicating a possible trouble spot in the grain bulk.**

## INVENTORY MANAGEMENT

Bin-level sensors help to estimate how much grain is contained in each pit, bin, or silo. Figure 7-5 displays a typical monitoring screen from a bin-level sensing system. Bin-level sensors help managers determine where to put new product coming in. Some level sensors use ultrasound; some are capacitive sensors, and some have mechanical moving parts much like trip switches or rotator switches. These level sensors also help prevent mistakes in selection of bins when placing grain in storage. If a bin-level sensor indicates a bin is already full, an automatic lock out will prevent more grain from entering the bin, thus preventing a catastrophic overflow or grain backup in a leg. A grain mismatch will also send an alarm to the operator and prevent mixing grain by mistake. Knowing the level of grain in a bin helps estimate inventory levels as well.

## EXAMPLE OF AUTOMATION IN A LARGE GRAIN FACILITY

Automation in a large on-farm or commercial grain facility simplifies the process of receiving trucks and grain, allows real-time monitoring of equipment and alarms, and makes the loadout procedure more accurate and more efficient.

### Truck Receiving

With automation, there is no need for a driver to exit the truck cab except to open truck gates at the receiving pit. The truck and its contents can be identified with bar codes or by intercom communication with the driver. After the loaded truck is weighed, the information is sent ahead

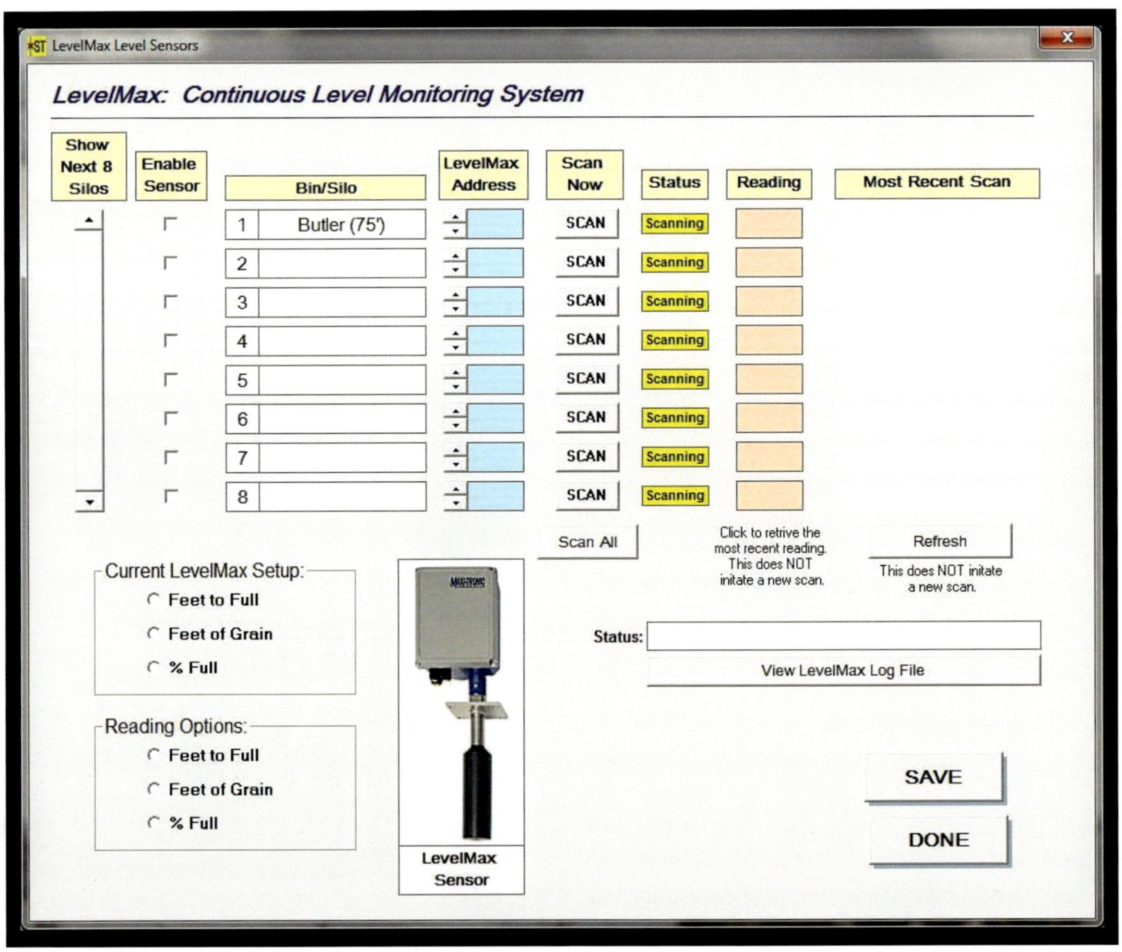


**FIGURE 7-5. Example of a monitoring screen for a bin-level sensing system.**

remotely to the outbound scale and to the grain accounting system. The truck driver receives a printout when the empty truck is weighed at the outbound scale (which offer the same scale as in-bound).

After weighing the incoming grain, a remote sampler collects grain and sends it pneumatically to the control room for inspection. After the grain is tested, the information is entered into software that calculates the USDA grade and records everything about the sample in the grain accounting system. All information goes to the accounting system, which maintains inventory information and has the ability to handle split accounts if necessary.

### Grain Receiving

The operator of the receiving pit can access the grade and test information via network access to the grain accounting system. The operator can select the destination bin, and information about grain level and accessibility of the targeted bin is displayed. Mistakes in bin selection are alerted and blocked by the software so that errors of dumping one type of grain on top of another are prevented. If the current load is a different type of grain than the previous load, a time delay allows the legs and conveyors to clear out before the grain change is made.

### Control Room

All operating equipment and bins as well as all alarm equipment for bearings, belts, and level switches can be monitored in real time. If there is a malfunction or an alert, a message is sent directly to the control room as well as cell phones. In some cases, all equipment involved and the feeding conveyors for that equipment go through a shut-down sequence to prevent devastating damage. The superintendent has complete visual monitoring capability of the entire grain facility complex in real time. All reports can be generated automatically, and accounts are updated automatically. Traceability is in place from receipt of the grain until the grain leaves the facility. Information about aeration and moisture control is recorded along with any other modifications to the grain or bin structure.

Maintenance supervisors also have a real-time view of all equipment. This includes reports of repairs, trouble alarms, and routine maintenance operations. Automation allows for better records and better response to problem areas before the problems grow into catastrophic events.

### Grain Loadout

Each truck or rail car has a radio frequency identification (RFID) tag. The grain accounting system knows the amount of grain that can be loaded in a specific truck or rail car. The system also determines the correct mixing and sourcing of the grain and controls the loading of the grain into the railcar or truck. Trucks and railcars can be loaded quickly and accurately with less risk to workers having to walk the line of rail cars checking to see the status at different points.

## SAFETY

Automation and control systems can be used to make a working environment safer. Heat and motion sensors installed on bearings and belts can alert maintenance personnel about rubbing of misaligned belts and hot bearings. This information is important so that maintenance can be performed before the malfunctioning unit causes a fire or dust explosion. Temperature monitoring in grain dryers can alert maintenance people of potential problems before a fire ignites. Also, power efficiency can be improved when belts are aligned correctly and bearings are not wearing beyond their service life. Belt rub blocks, shown in Figures 7-6 through 7-8, indicate that a belt is not aligned properly. Bearing temperature sensors like the one in Figure 7-9 indicate that bearings are wearing or malfunctioning

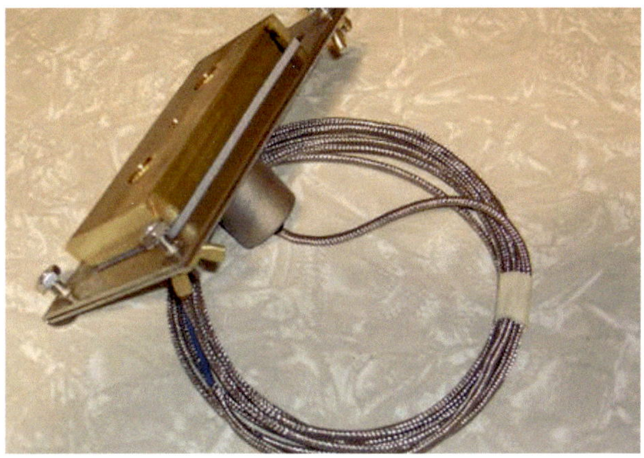

**Figure 7-6. Belt "rub" block to detect belt misalignment.**

**Figure 7-7. Worn belt alignment block.**

and causing heat to build. Belt speed monitors, pictured in Figures 7-10 and 7-11, indicate slippage of belts, and grain flow monitors indicate if flow of grain has stopped. In fact, the Occupational Safety and Health Administration's (OSHA) standard 1910.272 sets forth requirements that bucket elevators be shut down if belt speeds drop below a

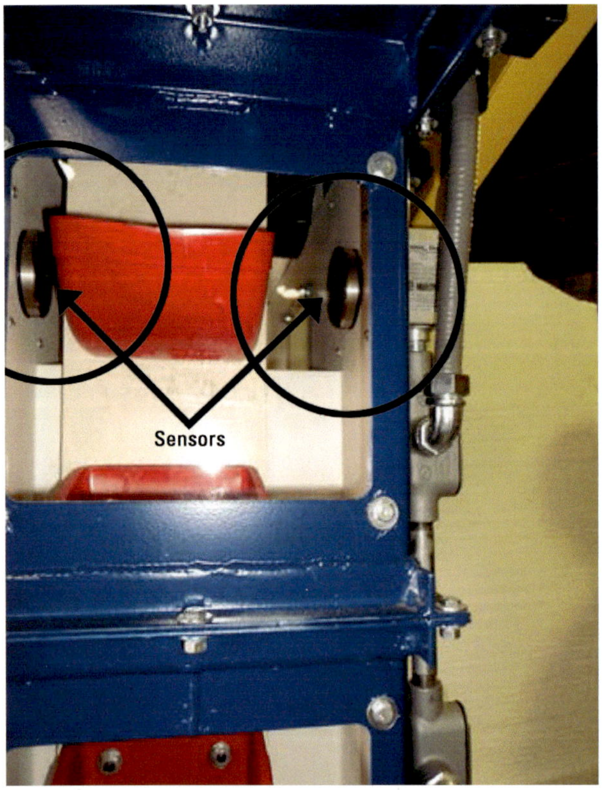


**Figure 7-8. Belt alignment sensors.**

**Figure 7-10. Bearing temperature sensor and shaft motion sensor.**

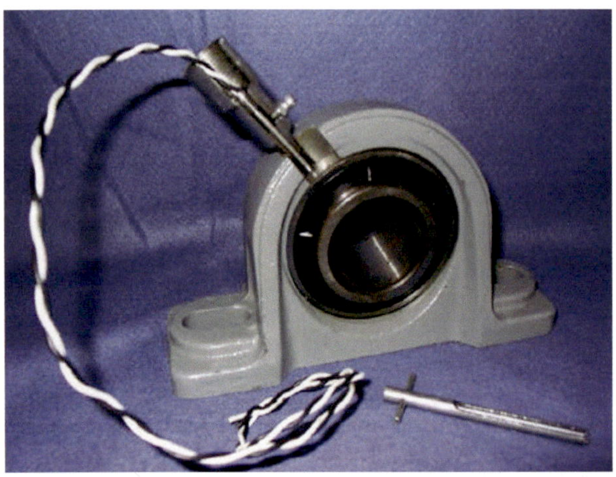

**Figure 7-9. Cutaway of a pillow block bearing temperature sensor.**

**Figure 7-11. Bearing temperature sensor and belt alignment sensor.**

percentage of the normal operating speed. All enclosures and electrical systems must be rated for the dusty environment inherent to grain handling facilities, certified by a Nationally Recognized Testing Laboratory (NRTL Class 2 Division 1 Group G for inside bucket elevators or belt conveyor or Division 2, Group G for outside a bucket elevator or conveyor). Figure 7-12 shows a bearing-temperature readout panel, and Figure 7-13 displays an example of a screen showing the temperatures of bucket-elevator bearings throughout a larger grain elevator. Maintenance managers can monitor all the facility's bucket elevator bearings from one central monitor. Figure 7-14 shows an example of historical bearing temperature data that can be used to diagnose problems or document maintenance improvements. Temperatures that continue to rise or are significantly different than other temperatures should be noted, and the cause of the temperature rise should be determined.

Figure 7-15, which illustrates a bucket elevator with its bearing and belt-alignment sensors, provides a good example of extensive use of monitoring systems for safety and performance.

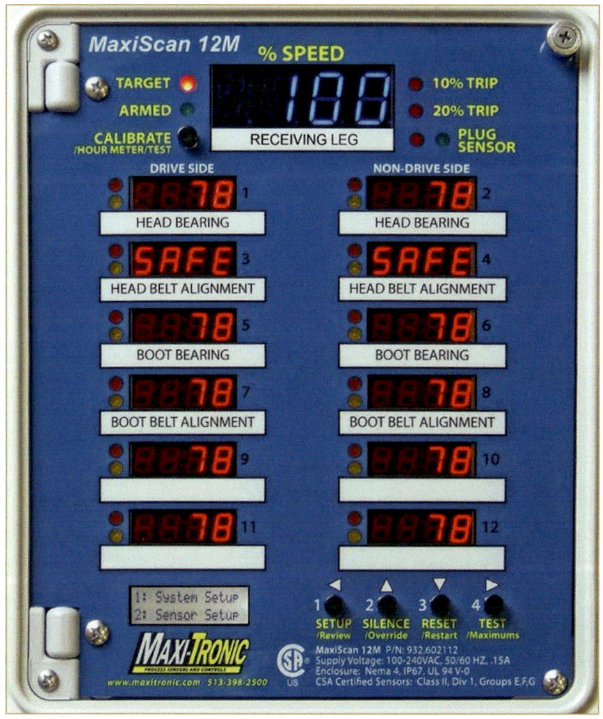


**Figure 7-12. Example of a bearing temperature and belt alignment readout panel.**

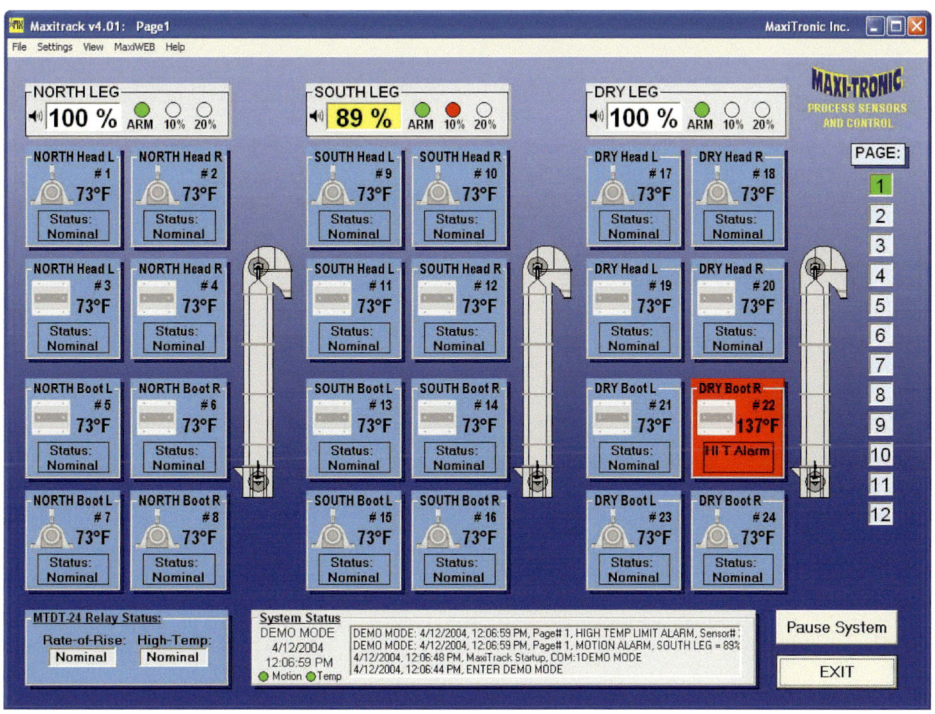


**Figure 7-13. Monitor screen for bearing temperature sensors.**

**FIGURE 7-14. Historical record of bearing temperatures.**
"C2" indicates a rise in bearing temperature, indicating that a problem exists and the bearing should be inspected.

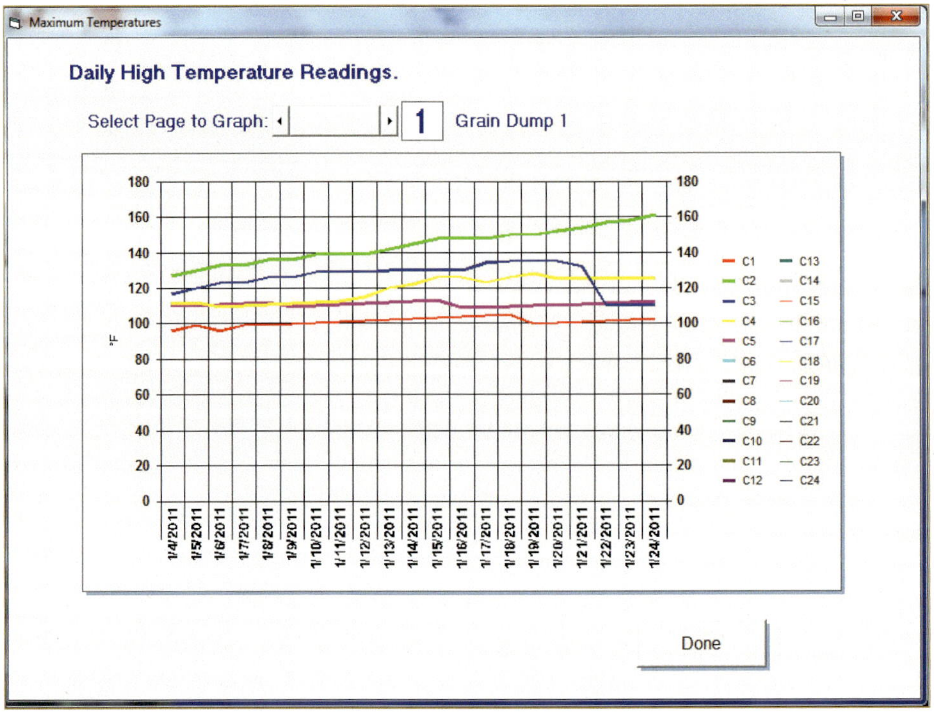


**FIGURE 7-15. Drawing of a bucket elevator system with critical monitoring points identified.**

*Photo courtesy Johnny Wheat (http://www.go4b.com/usa/technical-support/technical-papers/bucket-elevator-monitoring.pdf)*

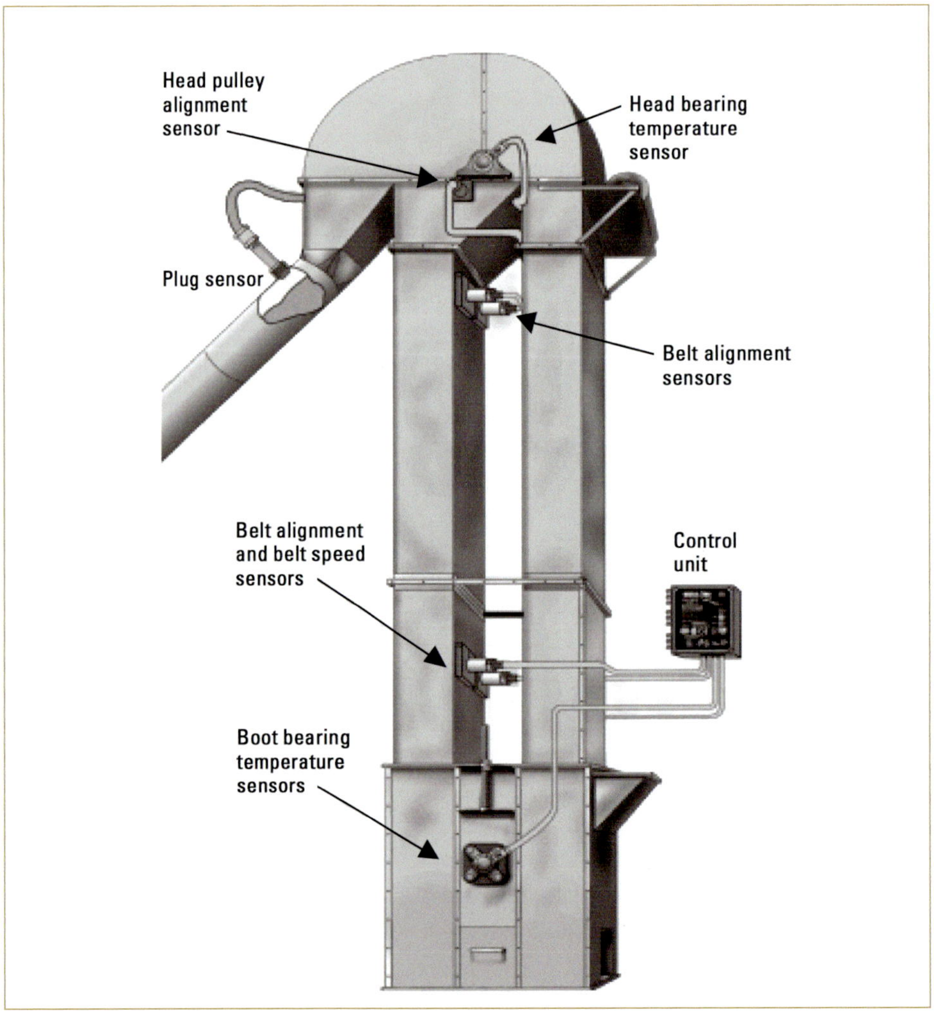

## COMPLETE MANAGEMENT SYSTEMS

All the controls and automated systems mentioned in this chapter can be integrated with a database that provides vital information for identity preservation, supply chain management, and product traceability. Such a system reads an identifier code when trucks deliver the grain to the grain facility. This identifier code can be indexed to a farmer and even to a specific production field. The weight, moisture content, quality inspection results, and receiving dump pit and bin identification information is then automatically stored in the database with the identifier code. An automatic accounting system results from having this information located in the same database. Producers know how much grain came from identified fields and have information about the quality of the grain and the grain's final destination to a customer. Information about how the grain was conditioned using aerator dryers and fans can be recorded automatically in the database using network technology.

## ADVANTAGES OF AUTOMATION

Controllers and automated systems have proven to be economical for on- and off-farm grain facilities by improving operational efficiency, by making the working environment safer, and by preventing mistakes. Automated systems also help reduce the labor required to manually sample and test conditions in bins and to turn aeration fans on and off. Because the controller is dedicated to the task, the grain is kept in better condition. Its environment is measured in real time, and adjustments are made by the aeration control system immediately, instead of when it fits into a worker's schedule to check and operate the systems.

While labor appears to be a big point of savings, it should be noted that expertise in handling and maintaining the electronics of these systems is required. This may require using outside contractor expertise or, if the facility is large enough, hiring an electronics and/or computer technician to help maintain and update the electronics and software required for intense automation. Consequently, the labor requirement may not be less, but it most certainly will require different skill sets. Training is also necessary for maintenance personnel and for workers that interface with the daily use of automated systems.

Working with a good sales and contracting company to establish a plan for automation has proven to be economical and a wise choice, particularly in the planning stages of any grain facility site. These contractors have experts on staff that know the advantages of designing an integrated system and can give good advice that matches the size and capacity of each operation. While automation and control systems are becoming a common part of grain handling, mistakes in planning can cause disaster. Work closely with a reputable contractor to get the most benefit from the change to automation or for installing automation in newly constructed facilities. Planning is key to successful integration and future expansion throughout the facility.

# APPENDIX

## CONVEYOR DISTANCES

**TABLE A-1. Inclined conveyor discharge heights and distances.**

Values are approximate. See Figure A-1 for variable definitions.

| | Conveyor angle | | | | | |
|---|---|---|---|---|---|---|
| | 25° | | 35° | | 45° | |
| L | H | D | H | D | H | H |
| ft | | | | | | |
| 21 | 9 | 19 | 12 | 17 | 15 | 15 |
| 26 | 11 | 23 ½ | 15 | 21 | 18 ½ | 18 ½ |
| 31 | 13 | 28 | 18 | 25 | 22 | 22 |
| 36 | 15 | 32 ½ | 21 | 29 | 25 ½ | 25 ½ |
| 41 | 17 ½ | 37 | 23 ½ | 33 ½ | 29 | 29 |
| 46 | 19 ½ | 41 ½ | 26 ½ | 37 ½ | 32 ½ | 32 ½ |
| 51 | 21 ½ | 46 | 29 ½ | 41 ½ | 36 | 36 |
| 56 | 23 ½ | 51 | 32 | 46 | 39 ½ | 39 ½ |
| 61 | 26 | 55 | 35 | 50 | 43 | 43 |
| 66 | 28 | 60 | 38 | 54 | 46 ½ | 46 ½ |
| 71 | 30 | 64 ½ | 40 ½ | 58 ½ | 50 | 50 |

**TABLE A-2. Minimum gravity spout angles and floor slopes.**

Values are absolute minimums for grain flow. Varnishing, waxing, or polishing wood and steel floors reduces friction and improves flow. If spouting is rusty, bulky, or damp, grain will flow better if spouting is scoured with dry hard-finish grain. If flow stops, it is difficult to restart.

| Material | Spout angle or floor slope | Slope equivalents rise:run |
|---|---|---|
| Grain, dry | 37° | ¾:1 |
| Grains, wet, hot, or tough | 45° | 1:1 |
| Meal | 60° | 1 ¾:1 |
| Pellets | 45° | 1:1 |
| Sunflower | 60° | 1 ¾:1 |

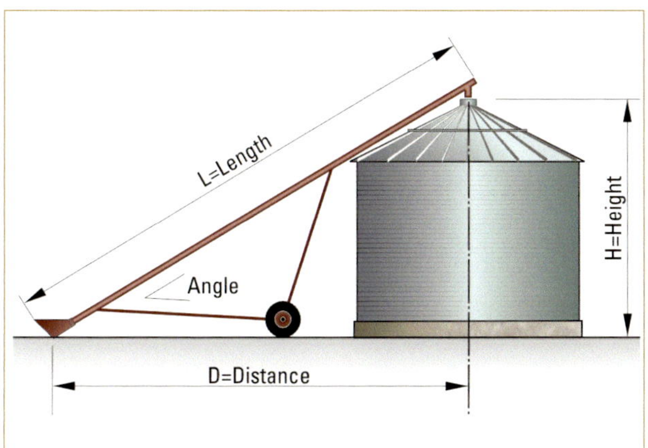


**FIGURE A-1. Inclined conveyor definition.**

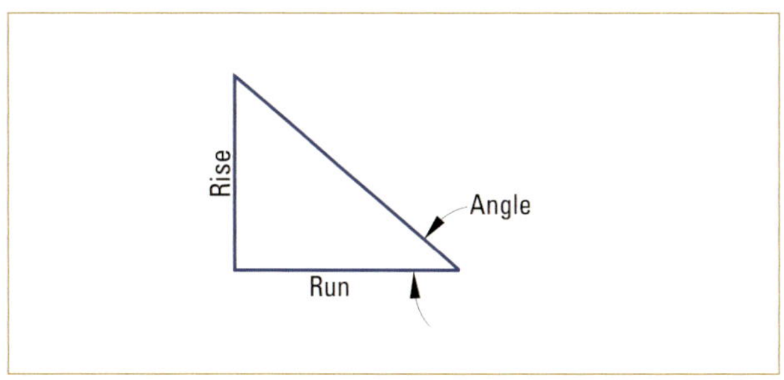


**FIGURE A-2. Slope triangle.**

**TABLE A-3. Gravity spout heights and lengths.**

See Figure A-3 for variable definitions.

| Distance to bin center | Spout angle | | | | | |
|---|---|---|---|---|---|---|
| | 37° | | 45° | | 60° | |
| D | L | H | L | H | L | H |
| | ft | | | | | |
| 16 | 20 | 12 | 23 | 16 | 32 | 28 |
| 18 | 22 ½ | 13 ½ | 25 ½ | 18 | 36 | 31 |
| 20 | 25 | 18 | 28 ½ | 20 | 40 | 35 |
| 22 | 27 ½ | 16 ½ | 31 | 22 | 44 | 38 |
| 24 | 30 | 18 | 34 | 24 | 48 | 42 |
| 26 | 32 ½ | 19 ½ | 37 | 26 | 52 | 45 |
| 28 | 35 | 21 | 40 | 28 | 56 | 48 ½ |
| 30 | 37 ½ | 22 ½ | 42 ½ | 30 | 60 | 52 |
| 35 | 44 | 26 ½ | 49 ½ | 35 | 70 | 61 |
| 40 | 50 | 30 | 57 | 40 | 80 | 70 |
| 45 | 57 | 34 | 64 | 45 | 90 | 78 |
| 50 | 63 | 38 | 71 | 50 | 100 | 87 |

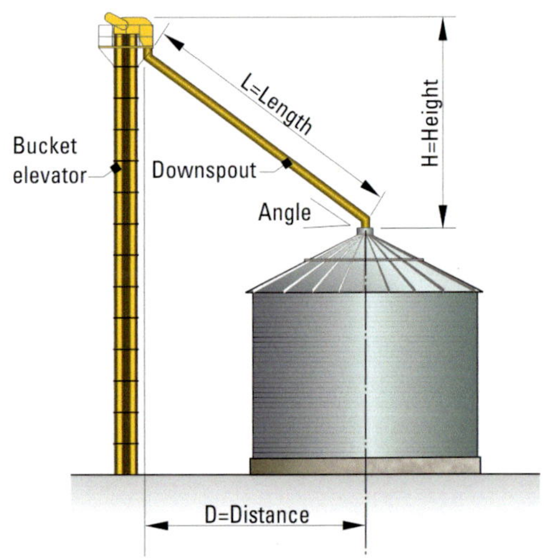


**FIGURE A-3. Gravity spout definition.**

## Inclined Conveyor Example

**EXAMPLE A-1.**

Determine the auger length and horizontal distance between an auger hopper and bin center if the vertical discharge height is 25 feet and the auger is inclined 35°. See Figure A-1.

**SOLUTION A-1.**

The vertical distance is the discharge height (H), horizontal distance is the distance between the auger hopper and bin hatch (D), and slope distance is the auger length (L). A 44-foot long auger is needed, and the hopper end is 36 feet from the center of the bin. See Figure A-4.

## Gravity Spout Examples

**EXAMPLE A-2.**

Determine the elevator discharge height and spout length for a gravity spout at a 60° incline discharging to a bin 20 feet from the bucket elevator.
See Figure A-3.

**SOLUTION A-2.**

1. For gravity spouting, vertical distance is the discharge height above the bin top (H); horizontal distance is the distance between the bucket elevator and bin top, (D) and slope distance is the spout length (L).

2. Using Table A-3, the spout angle = 60° and the distance to bin center D = 20 feet.

   The discharge height above the bin top is 35 feet and the spout length is 40 feet.

3. Using Figure A-4, the elevator discharge must be 35 feet above the bin top, and the spout is 40 feet long.

## EXAMPLE A-3.

Determine the vertical discharge height for a gravity spout at a 37° incline discharging dry grain from a bucket elevator to a bin top 24 feet away.

## SOLUTION A-3.

1. Using Figure A-2, the slope equivalent (rise:run) is ¾:1. For this case, run = 24 feet; determine rise

$$\text{Rise} = \tfrac{3}{4} \times \text{Run} = \tfrac{3}{4} \times 24 = 18 \text{ feet}$$

Vertical discharge height between the bucket elevator discharge and bin top is 18 feet.

2. Using Table A-3, spout angle 37° and distance to bin center, D = 24 feet, determine the discharge height (H).

From Table A-3, the discharge height is 18 feet. The length of the conveyor, L = 25 feet.

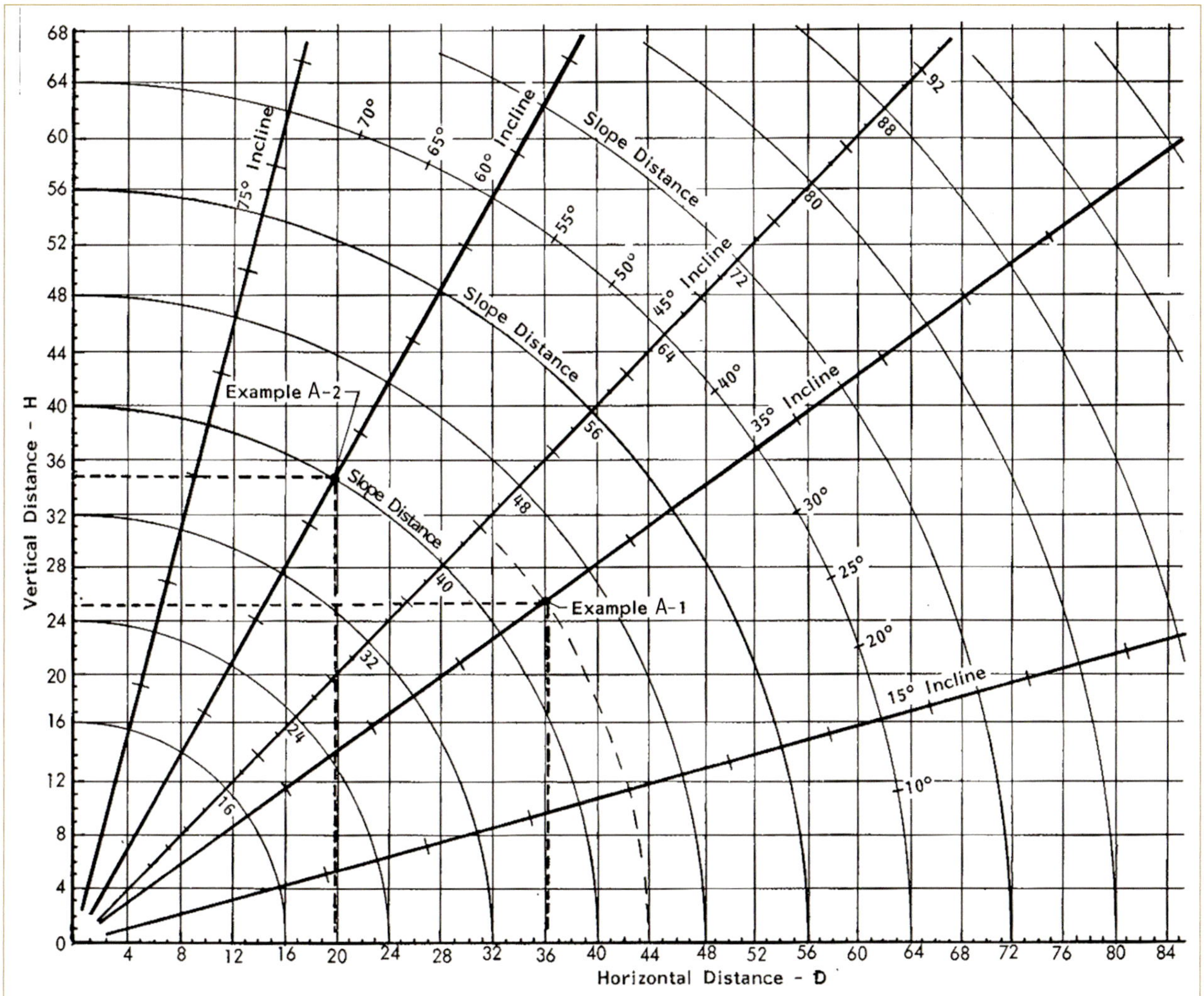


**FIGURE A-4. Slope chart for conveyors and gravity spouts.**
Chart variables are vertical distance or rise (V); horizontal distance or run (H); and conveyor incline (I). Use any two variables to determine the other two. For example, if conveyor length and slope are known, then vertical and horizontal distance can be determined.

# STORAGE VOLUME TABLES

**TABLE A-4. Grain seed densities.**

Based on weights and measures used by the U.S. Department of Agriculture. Adapted from ASAE Data D241.2. 1984

| Grain or seed | Unit | Approximate net weight[a] lb | Bulk density $lb/ft^3$ |
|---|---|---|---|
| Alfalfa | Bushel | 60 | 48.0 |
| Barley | Bushel | 48 | 38.4 |
| Beans | | | |
| Lima, dry | Bushel | 56 | 44.8 |
| Lima, unshelled | Bushel | 32 | 25.6 |
| Snap | Bushel | 30 | 24.0 |
| Other, dry (navy pinto, etc.) | Bushel | 60 | 48.0 |
| Bluegrass | Bushel | 14-30 | 11.2-24.0 |
| Broomcorn | Bushel | 44-50 | 35.2-40.0 |
| Buckwheat | Bushel | 48-52 | 38.4-41.6 |
| Castor beans | Bushel | 46 | 36.8 |
| Clover | Bushel | 60 | 48.0 |
| Corn | | | |
| Ear, husked | Bushel | 70[b] | 28.0 |
| Shelled | Bushel | 56 | 44.8 |
| Green sweet | Bushel | 35 | 28.0 |
| Cottonseed | Bushel | 32 | 25.6 |
| Cowpeas | Bushel | 60 | 48.0 |
| Flaxseed | Bushel | 56 | 44.8 |
| Grain sorghums | Bushel | 56 & 50 | 44.8 & 40.0 |
| Hempseed | Bushel | 44 | 35.2 |
| Hungarian millet | Bushel | 48 & 50 | 38.4 & 40.0 |
| Kafir | Bushel | 56 & 50 | 44.8 & 40.0 |
| Kapok | Bushel | 35-40 | 28.0-32.0 |
| Lentils | Bushel | 60 | 48.0 |
| Millet | Bushel | 48-50 | 38.4-40.0 |
| Mustard | Bushel | 58-60 | 46.4-48.0 |
| Oats | Bushel | 32 | 25.6 |
| Orchard grass | Bushel | 14 | 11.2 |

| Grain or seed | Unit | Approximate net weight[a] lb | Bulk density $lb/ft^3$ |
|---|---|---|---|
| Peanuts, unshelled | | | |
| Virginia type | Bushel | 17 | 13.6 |
| Runners, Southeastern | Bushel | 21 | 16.8 |
| Spanish peanuts | | | |
| Southeastern | Bushel | 25 | 19.7 |
| Southwestern | Bushel | 25 | 19.8 |
| Perilla | Bushel | 37-40 | 29.6-32.0 |
| Popcorn | | | |
| On ear | Bushel | 70 | 28.0 |
| Shelled | Bushel | 56 | 44.8 |
| Poppy | Bushel | 46 | 36.8 |
| Rapeseed | Bushel | 50 & 60 | 40.0 & 48.0 |
| Redtop | Bushel | 50 & 60 | 40.0 & 48.0 |
| Rice, rough | Bushel | 45 | 36.0 |
| Rye | Bushel | 56 | 44.8 |
| Sesame | Bushel | 46 | 36.8 |
| Sorgo | Bushel | 50 | 40.0 |
| Soybeans | Bushel | 60 | 48.0 |
| Spelt (p. wheat) | Bushel | 40 | 32.0 |
| Sudan grass | Bushel | 40 | 32.0 |
| Sunflower | | | |
| Non-oil | Bushel | 24 | 19.2 |
| Oil | Bushel | 32 | 25.6 |
| Timothy | Bushel | 45 | 36.0 |
| Velvet beans (hulled) | Bushel | 60 | 48.0 |
| Vetch | Bushel | 60 | 48.0 |
| Wheat | Bushel | 60 | 48.0 |

[a] Net weight is the standard bushel weight if one has been defined.

[b] 70 lb of husked, ear corn yields 1 bu, or 56 lb of shelled corn. 70 lb ear corn occupies 2 volume bushels (2.5 $ft^3$).

## TABLE A-5. Storage capacity of level-full round bins.

Based on 1.245 cu ft = 1 bushel. Capacities vary by material density, packing and shape of grain surface.

| Grain depth, ft | Bin diameter, ft | | | | | | | | | | | | |
|---|---|---|---|---|---|---|---|---|---|---|---|---|---|
| | 18 | 21 | 24 | 27 | 30 | 33 | 36 | 42 | 48 | 60 | 72 | 90 | 105 |
| | volume bushels | | | | | | | | | | | | |
| 1 | 204 | 278 | 363 | 460 | 568 | 687 | 818 | 1,113 | 1,453 | 2,271 | 3,270 | 5,110 | 6,955 |
| 4 | 818 | 1,113 | 1,453 | 1,840 | 2,271 | 2,748 | 3,270 | 4,451 | 5,814 | 9,084 | 13,081 | 20,439 | 27,820 |
| 6 | 1,226 | 1,669 | 2,180 | 2,759 | 3,407 | 4,122 | 4,905 | 6,677 | 8,721 | 13,626 | 19,622 | 30,659 | 41,730 |
| 8 | 1,635 | 2,226 | 2,907 | 3,679 | 4,542 | 5,496 | 6,541 | 8,902 | 11,628 | 18,168 | 26,162 | 40,878 | 55,640 |
| 10 | 2,044 | 2,782 | 3,634 | 4,599 | 5,678 | 6,870 | 8,176 | 11,128 | 14,534 | 22,710 | 32,703 | 51,098 | 69,550 |
| 12 | 2,453 | 3,338 | 4,360 | 5,519 | 6,813 | 8,244 | 9,811 | 13,354 | 17,441 | 27,252 | 39,243 | 61,317 | 83,460 |
| 14 | 2,861 | 3,895 | 5,087 | 6,438 | 7,949 | 9,618 | 11,446 | 15,579 | 20,348 | 31,794 | 45,784 | 71,537 | 97,369 |
| 16 | 3,270 | 4,451 | 5,814 | 7,358 | 9,084 | 10,992 | 13,081 | 17,805 | 23,255 | 36,336 | 52,324 | 81,756 | 111,279 |
| 18 | 3,679 | 5,008 | 6,541 | 8,278 | 10,220 | 12,366 | 14,716 | 20,030 | 26,162 | 40,878 | 58,865 | 91,976 | 125,189 |
| 20 | 4,088 | 5,564 | 7,267 | 9,198 | 11,355 | 13,740 | 16,351 | 22,256 | 29,069 | 45,420 | 65,405 | 102,195 | 139,099 |
| 22 | 4,497 | 6,120 | 7,994 | 10,117 | 12,491 | 15,114 | 17,986 | 24,481 | 31,976 | 49,962 | 71,946 | 112,415 | 153,009 |
| 24 | 4,905 | 6,677 | 8,721 | 11,037 | 13,626 | 16,488 | 19,622 | 26,707 | 34,883 | 54,504 | 78,486 | 122,634 | 166,919 |
| 26 | 5,314 | 7,233 | 9,447 | 11,957 | 14,762 | 17,861 | 21,257 | 28,933 | 37,790 | 59,046 | 85,027 | 132,854 | 180,829 |
| 28 | 5,723 | 7,790 | 10,174 | 12,877 | 15,897 | 19,235 | 22,892 | 31,158 | 40,696 | 63,588 | 91,567 | 143,073 | 194,739 |
| 30 | 6,132 | 8,346 | 10,901 | 13,796 | 17,033 | 20,609 | 24,527 | 33,387 | 43,603 | 68,130 | 98,108 | 153,293 | 208,649 |
| 32 | 6,541 | 8,902 | 11,628 | 14,716 | 18,168 | 21,983 | 26,162 | 35,609 | 46,510 | 72,672 | 104,648 | 163,513 | 222,559 |
| 34 | 6,949 | 9,459 | 12,354 | 15,636 | 19,304 | 23,357 | 27,797 | 37,835 | 49,417 | 77,214 | 111,189 | 173,732 | 236,469 |
| 36 | 7,358 | 10,015 | 13,081 | 16,556 | 20,439 | 24,731 | 29,432 | 40,061 | 52,324 | 81,756 | 117,729 | 183,952 | 250,379 |
| 38 | 7,767 | 10,572 | 13,808 | 17,475 | 21,575 | 26,105 | 31,067 | 42,286 | 55,231 | 86,298 | 124,270 | 194,171 | 264,288 |
| 40 | 8,176 | 11,128 | 14,534 | 18,395 | 22,710 | 27,479 | 32,703 | 44,512 | 58,138 | 90,840 | 130,810 | 204,391 | 278,198 |
| 42 | 8,584 | 11,684 | 15,261 | 19,315 | 23,846 | 28,853 | 34,338 | 46,737 | 61,045 | 95,382 | 137,351 | 214,610 | 292,108 |
| 44 | 8,993 | 12,241 | 15,988 | 20,235 | 24,981 | 30,227 | 35,973 | 48,963 | 63,952 | 99,924 | 143,891 | 224,830 | 306,018 |
| 46 | 9,402 | 12,797 | 16,715 | 21,154 | 26,117 | 31,601 | 37,608 | 51,189 | 66,858 | 104,466 | 150,432 | 235,049 | 319,928 |
| 48 | 9,811 | 13,354 | 17,441 | 22,074 | 27,252 | 32,975 | 39,243 | 53,414 | 69,765 | 109,008 | 156,972 | 245,269 | 333,838 |
| 50 | 10,220 | 13,910 | 18,168 | 22,994 | 28,388 | 34,349 | 40,878 | 55,640 | 72,672 | 113,550 | 163,513 | 225,488 | 347,748 |
| 52 | 10,628 | 14,466 | 18,895 | 23,914 | 29,523 | 35,723 | 42,513 | 57,865 | 75,579 | 118,092 | 170,053 | 265,708 | 361,658 |
| 54 | 11,037 | 15,023 | 19,622 | 24,833 | 30,659 | 37,097 | 44,148 | 60,091 | 78,486 | 122,634 | 176,594 | 275,927 | 375,568 |
| 56 | 11,446 | 15,579 | 20,348 | 25,753 | 31,794 | 38,471 | 45,784 | 62,316 | 81,393 | 127,176 | 183,134 | 286,147 | 389,478 |
| 58 | 11,855 | 16,136 | 21,075 | 26,673 | 32,930 | 39,845 | 47,419 | 64,542 | 84,300 | 131,718 | 189,675 | 296,366 | 403,388 |
| 60 | 12,263 | 16,692 | 21,802 | 27,593 | 34,065 | 41,219 | 49,054 | 66,768 | 87,207 | 136,260 | 196,215 | 306,586 | 417,298 |

The information in this table is also available as Spreadsheet S 5-1.

## TABLE A-6. Storage capacity of full cones. Add to capacity of level-fill round bin to estimate total of a peaked grain mass.

Based on 1.245 cu ft = 1 bushel. Capacities vary by material density, packing and shape of grain surface.

| | Angle of repose | Bin diameter, ft | | | | | | | | | | | | | |
|---|---|---|---|---|---|---|---|---|---|---|---|---|---|---|---|
| | | 18 | 21 | 24 | 27 | 30 | 33 | 36 | 42 | 48 | 60 | 72 | 90 | 105 | Factor |
| Corn | 23 | 260 | 413 | 617 | 878 | 1,205 | 1,604 | 2,082 | 3,306 | 4,936 | 9,640 | 16,658 | 32,535 | 51,664 | 0.4245 |
| Soybeans | 25 | 286 | 454 | 678 | 965 | 1,324 | 1,762 | 2,287 | 3,632 | 5,422 | 10,590 | 18,299 | 35,741 | 56,755 | 0.4663 |
| Wheat | 29 | 340 | 540 | 806 | 1,147 | 1,574 | 2,094 | 2,719 | 4,318 | 6,445 | 12,588 | 21,753 | 42,486 | 67,466 | 0.5543 |

The information in this table is also available as Spreadsheet S 5-1.

| TABLE A-7. Storage capacity of flat storage. | | | | | | | | |
|---|---|---|---|---|---|---|---|---|
| Based on 1.245 cu ft = 1 bushel.<br>Capacities vary by material density, packing and shape of grain surface. | | | | | | | | |
| **Building** | | **Grain height** | | **Building capacity** | | | | **Total airflow cfm bu** |
| **Width ft** | **Length ft** | **Wall ft** | **Peak ft** | **Ends bu** | **Middle bu** | **Base bu** | **total bu** | |
| 40 | 40 | 2 | 10.5 | 2,856 | 0 | 2,570 | 5,427 | 10,085 |
| 40 | 40 | 4 | 12.5 | 2,856 | 0 | 5,141 | 7,997 | 1,599 |
| 40 | 100 | 2 | 10.5 | 2,856 | 8,183 | 6,426 | 17,465 | 3,493 |
| 40 | 100 | 6 | 14.5 | 2,856 | 8,183 | 19,277 | 30,316 | 6,063 |

The information in this table is also available as Spreadsheet S 5-2. See Figure 5-3.

| TABLE A-8. Storage capacity for dry grain in level-fill tower silos. | | | | | | | |
|---|---|---|---|---|---|---|---|
| Based on 1.245 cu ft = 1 bushel.<br>Capacities vary by packing, shape of grain surface, and volume of aeration tubes. | | | | | | | |
| **Grain depth ft** | **Silo diameter ft** | | | | | | |
| | **12** | **14** | **16** | **18** | **20** | **24** | **30** |
| 1 | 94 | 124 | 161 | 204 | 252 | 363 | 568 |
| 5 | 454 | 618 | 807 | 1,022 | 1,262 | 1,817 | 2,839 |
| 10 | 908 | 1,236 | 1,615 | 2,044 | 2,523 | 3,634 | 5,678 |
| 15 | 1,363 | 1,855 | 2,422 | 3,066 | 3,785 | 5,450 | 8,516 |
| 20 | 1,817 | 2,473 | 3,230 | 4,088 | 5,047 | 7,267 | 11,355 |
| 25 | 2,271 | 3,091 | 4,037 | 5,110 | 6,308 | 9,084 | 14,194 |
| 30 | 2,725 | 3,709 | 4,845 | 6,132 | 7,570 | 10,901 | 17,033 |
| 35 | 3,179 | 4,328 | 5,652 | 7,154 | 8,832 | 12,718 | 19,871 |
| 40 | 3,634 | 4,946 | 6,460 | 8,176 | 10,093 | 14,534 | 22,710 |
| 45 | 4,088 | 5,564 | 7,267 | 9,198 | 11,355 | 16,351 | 25,549 |
| 50 | 4,542 | 6,182 | 8,075 | 10,220 | 12,617 | 18,168 | 28,388 |
| 55 | | 6,800 | 8,882 | 11,241 | 13,878 | 19,985 | 31,226 |
| 60 | | | 9,690 | 12,263 | 15,140 | 21,802 | 34,065 |
| 65 | | | 10,497 | 13,285 | 16,402 | 23,618 | 36,904 |
| 70 | | | | 14,307 | 17,663 | 25,435 | 39,743 |
| 75 | | | | | 18,925 | 27,252 | 42,581 |
| 80 | | | | | 20,187 | 29,069 | 45,420 |
| 85 | | | | | | 30,886 | 48,259 |
| 90 | | | | | | 32,703 | 51,098 |
| 95 | | | | | | | 53,936 |
| 100 | | | | | | | 56,775 |
| 105 | | | | | | | 59,614 |
| 110 | | | | | | | 62,453 |

The information in this table is also available as Spreadsheet S 5-3.

**TABLE A-9. Storage capacity above level-fill for dry grain in tower silos. Add to capacity of level-fill tower silo to estimate total of a peaked grain mass.**

Based on 1.245 cu ft = 1 bushel.
Capacities vary by packing, shape of grain surface, and volume of aeration tubes.

| Grain | Angle of repose, degrees | Silo diameter, ft | | | | | | | |
|---|---|---|---|---|---|---|---|---|---|
| | | 12 | 14 | 16 | 18 | 20 | 24 | 30 | Factor |
| Corn | 23 | 77 | 122 | 183 | 260 | 357 | 617 | 1,205 | 0.424 |
| Soybeans | 25 | 85 | 135 | 201 | 386 | 392 | 678 | 1,324 | 0.466 |
| Wheat | 29 | 101 | 160 | 239 | 340 | 466 | 806 | 1,574 | 0.554 |

The information in this table is also available as Spreadsheet S 5-3.

**TABLE A-10. Select storage capacities of outdoor grain piles containing corn aerated at 0.2 cfm/bu.**

Based on 1.245 cu ft = 1 bushel. Capacities vary by material density, packing and shape of grain surface.

| Diameter (D) ft | Wall height (H) ft | Peak height (H) ft | Capacity bu | Airflow cfm |
|---|---|---|---|---|
| 40 | 4 | 12.5 | 6,894 | 1,379 |
| 90 | 5 | 24.1 | 58,084 | 11,617 |
| 150 | 4 | 35.8 | 207,400 | 41,480 |
| 300 | 6 | 69.7 | 1,544,656 | 308,931 |

The information in this table is also available as Spreadsheet S 5-4.

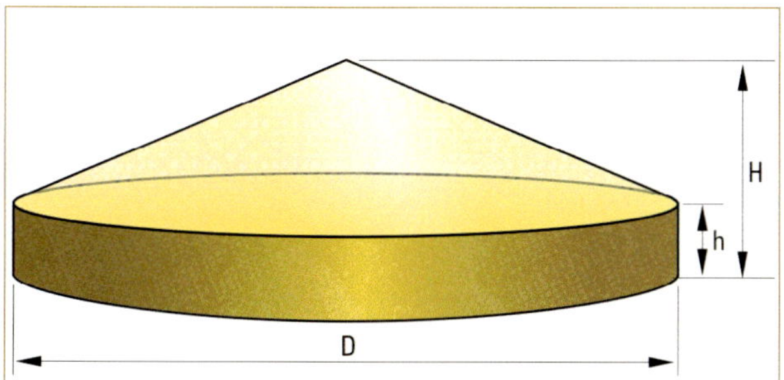


**TABLE A-11. Grain storage capacity of large (silo-type) horizontal plastic grain bags (silo bags).**

Based on 1.245 cu ft = 1 bushel. Capacities vary by material density, packing and shape at end of the bag.

| Bag length ft | Bag diameter, ft | | | |
|---|---|---|---|---|
| | 6 | 8 | 10 | 12 |
| 50 | 1,136 | 2,019 | 3,154 | 4,542 |
| 100 | 2,271 | 7,037 | 6,308 | 9,084 |
| 150 | 3,407 | 6,056 | 9,463 | 13,626 |
| 200 | 4,542 | 8,075 | 12,617 | 18,168 |
| 300 | 6,813 | 12,112 | 18,925 | 27,252 |

The information in this table is also available as Spreadsheet S 5-5.

**TABLE A-12. Storage volumes for bulk materials.**

Approximate data, based primarily on California Agricultural Circular 517.

| Material | lb/ft$^3$ | ft$^3$/ton |
|---|---|---|
| Alfalfa meal | 15 | 134 |
| Alfalfa, chopped | 12 | 167 |
| Barley meal | 28 | 72 |
| Beet pulp, dried | 15 | 134 |
| Bran | 16 | 125 |
| Brewers' grains, dry | 15 | 134 |
| Buckwheat bran | 15 | 134 |
| Buckwheat middlings | 23 | 87 |
| Coconut meal | 38 | 53 |
| Concentrates, typical | 45 | 45 |
| Corn meal | 38 | 53 |
| Ground ear corn, dry | 36 | 56 |
| Cotton seed meal | 38 | 53 |
| Distillers' grain, dry | 15 | 134 |
| Gluten feed | 33 | 61 |
| Gluten meal | 46 | 44 |
| Hominy meal | 28 | 72 |
| Kafir meal | 27 | 74 |
| Lime | 60 | 33 |
| Linseed meal | 23 | 87 |
| Malt sprouts | 15 | 134 |
| Mixed mill feed (bran & middlings) | 15 | 134 |
| Molasses | 78 | 26 |
| Molasses beet pulp | 20 | 100 |
| Meat scraps | 34 | 59 |
| Oats, ground | 20 | 100 |
| Rice bran | 20 | 100 |
| Salt, fine | 50 | 40 |
| Soybean meal | 42 | 48 |
| Tankage | 32 | 63 |
| Wheat bran | 14 | 143 |
| Wheat feed, mixed | 15 | 134 |
| Wheat, ground | 45 | 45 |
| Wheat middling (Std) | 20 | 100 |
| Wheat screenings | 27 | 74 |
| Pellets, mixed feed | 40 | 50 |
| Pellets, ground hay | 45 | 45 |

**TABLE A-13. Capacities of hopper-bottom round bulk bins.**

Approximate values. Overall height is the distance from ground to roof top and includes 2 feet of clearance between ground and hopper bottom. Roof at 35° angle. Bin volume includes roof volume.

| Bin diameter ft | Hopper angle degrees | Overall height ft | Total ft³ | bu | Volume/ft of height* ft³ | Volume/ft of height* bu |
|---|---|---|---|---|---|---|
| **Sidedraw** | | | | | | |
| 6 | 60 | 15 | 227 | 182 | 28 | 23 |
| | | 20 | 368 | 295 | | |
| | | 25 | 510 | 410 | | |
| 9 | 60 | 22 | 760 | 611 | 64 | 51 |
| | | 26 | 1,015 | 815 | | |
| | | 30 | 1,270 | 1,109 | | |
| **Centerdraw** | | | | | | |
| 6 | 60 | 10 | 145 | 117 | 28 | 23 |
| | | 15 | 286 | 230 | | |
| | | 20 | 428 | 344 | | |
| 9 | 60 | 18 | 681 | 547 | 64 | 51 |
| | | 22 | 935 | 751 | | |
| | | 26 | 1,190 | 955 | | |
| 12 | 60 | 24 | 1,613 | 1,296 | 113 | 91 |
| | | 30 | 2,292 | 1,841 | | |
| | | 36 | 2,970 | 2,386 | | |
| 12 | 45 | 20 | 1,493 | 1,199 | 113 | 91 |
| | | 30 | 2,624 | 2,108 | | |
| | | 40 | 3,755 | 3,016 | | |
| 15 | 45 | 24 | 2,739 | 2,200 | 177 | 142 |
| | | 36 | 4,860 | 3,903 | | |
| | | 48 | 6,980 | 5,607 | | |
| 18 | 45 | 24 | 3,512 | 2,821 | 255 | 204 |
| | | 36 | 5,268 | 4,231 | | |
| | | 48 | 9,619 | 7,726 | | |
| 21 | 45 | 24 | 4,191 | 3,366 | 346 | 278 |
| | | 36 | 8,348 | 6,705 | | |
| | | 48 | 12,504 | 10,043 | | |

*For calculating volume of different height bins, add or subtract from table value.

GRAIN MOISTURE RELATIONSHIPS ▶

## TABLE A-14. Weight of grain to yield a standard bushel.

A bushel is the amount of grain required to yield: 56 lb of shelled corn, 14.0, 15.0, or 15.5% moisture content (m.c.); 60 lb of wheat, 13.5% m.c.; 60 lb of soybeans, 13.0% m.c.; 48 lb of barley, 14.5% m.c.; 32 lb of oats, 14.0% m.c.; and 56 lb of rye, 14.0 % m.c. Sunflower and grain sorghum are based on 100 lb (CWT) rather than bushels. For sunflower and grain sorghum, table values are the amount of grain to yield 100 lb sunflower at 10% m.c. and 100 lb of grain sorghum at 14% m.c.

lb. dry matter/bu = 47.32 for corn, 51.90 for wheat, 52.20 for soybeans, 41.04 for barley, 27.52 for oats, and 48.16 for rye.
lb dry matter/cwt = 90 for sunflower and 86 for grain sorghum.

| Moisture content, % | Shelled corn | Shelled corn | Shelled corn | Wheat | Soybeans | Barley | Oats | Rye | Sunflower | Grain sorghum |
|---|---|---|---|---|---|---|---|---|---|---|
| | | | lb/bu | | | | | | lb/cwt base | |
| 5.0 | 50.69 | 50.11 | 49.81 | 54.63 | 54.95 | 43.20 | 28.97 | 50.69 | 94.74 | 90.53 |
| 5.5 | 50.96 | 50.37 | 50.07 | 54.92 | 55.24 | 43.43 | 29.12 | 50.96 | 95.24 | 91.01 |
| 6.0 | 51.23 | 50.64 | 50.34 | 55.21 | 55.53 | 43.66 | 29.28 | 51.23 | 95.74 | 91.49 |
| 6.5 | 51.51 | 50.91 | 50.61 | 55.51 | 55.83 | 43.89 | 29.43 | 51.51 | 96.26 | 91.98 |
| 7.0 | 51.78 | 51.18 | 50.88 | 55.81 | 56.13 | 44.13 | 29.59 | 51.79 | 96.78 | 92.47 |
| 7.5 | 52.06 | 51.46 | 51.16 | 56.11 | 56.43 | 44.37 | 29.75 | 52.07 | 97.30 | 92.97 |
| 8.0 | 52.35 | 51.74 | 51.44 | 56.42 | 56.74 | 44.61 | 29.91 | 52.35 | 97.83 | 93.48 |
| 8.5 | 52.63 | 52.02 | 51.72 | 56.72 | 57.05 | 44.85 | 30.08 | 52.63 | 98.36 | 93.99 |
| 9.0 | 52.92 | 52.31 | 52.00 | 57.03 | 57.36 | 45.10 | 30.24 | 52.92 | 98.90 | 94.51 |
| 9.5 | 53.22 | 52.60 | 52.29 | 57.35 | 57.68 | 45.35 | 30.41 | 53.22 | 99.45 | 95.03 |
| 10.0 | 53.51 | 52.89 | 52.58 | 57.67 | 58.00 | 45.60 | 30.58 | 53.51 | **100.00** | 95.56 |
| 10.5 | 53.81 | 53.18 | 52.87 | 57.99 | 58.32 | 45.85 | 30.75 | 53.81 | 100.56 | 96.09 |
| 11.0 | 54.11 | 53.48 | 53.17 | 58.31 | 58.65 | 46.11 | 30.92 | 54.11 | 101.12 | 96.63 |
| 11.5 | 54.42 | 53.79 | 53.47 | 58.64 | 58.98 | 46.37 | 31.10 | 54.42 | 101.69 | 97.18 |
| 12.0 | 54.73 | 54.09 | 53.77 | 58.97 | 59.32 | 46.64 | 31.28 | 54.73 | 102.28 | 97.73 |
| 12.5 | 55.04 | 54.40 | 54.08 | 59.31 | 59.65 | 46.90 | 31.45 | 55.04 | 102.86 | 98.29 |
| 13.0 | 55.36 | 54.71 | 54.39 | 59.65 | **60.00** | 47.17 | 31.63 | 55.36 | 103.46 | 98.85 |
| 13.5 | 55.68 | 55.03 | 54.71 | **60.00** | 60.35 | 47.45 | 31.81 | 55.68 | 104.05 | 99.42 |
| 14.0 | **56.00** | 55.35 | 55.02 | 60.35 | 60.70 | 47.72 | **32.00** | **56.00** | 104.66 | **100.00** |
| 14.5 | 56.33 | 55.67 | 55.35 | 60.70 | 61.06 | **48.00** | 32.19 | 56.33 | 105.26 | 100.58 |
| 15.0 | 56.66 | **56.00** | 55.67 | 61.06 | 61.41 | 48.28 | 32.38 | 56.66 | 105.89 | 101.18 |
| 15.5 | 56.99 | 56.33 | **56.00** | 61.43 | 61.78 | 48.57 | 32.57 | 57.00 | 106.52 | 101.78 |
| 16.0 | 57.33 | 56.67 | 56.33 | 61.79 | 62.14 | 48.86 | 32.76 | 57.34 | 107.15 | 102.38 |
| 16.5 | 57.68 | 57.01 | 56.67 | 62.16 | 62.52 | 49.15 | 32.96 | 57.68 | 107.79 | 102.99 |
| 17.0 | 58.02 | 57.35 | 57.01 | 62.53 | 62.89 | 49.45 | 33.16 | 58.03 | 108.44 | 103.61 |
| 17.5 | 58.38 | 57.70 | 57.36 | 62.91 | 63.28 | 49.75 | 33.36 | 58.38 | 109.10 | 104.24 |
| 18.0 | 58.73 | 58.05 | 57.71 | 63.30 | 63.66 | 50.05 | 33.56 | 58.74 | 109.76 | 104.88 |
| 18.5 | 59.09 | 58.40 | 58.06 | 63.69 | 64.05 | 50.36 | 33.77 | 59.10 | 110.44 | 105.52 |
| 19.0 | 59.46 | 58.77 | 58.42 | 64.08 | 64.45 | 50.67 | 33.98 | 59.46 | 111.11 | 106.17 |
| 19.5 | 59.83 | 59.13 | 58.78 | 64.47 | 64.85 | 50.98 | 34.18 | 59.83 | 111.81 | 106.83 |
| 20.0 | 60.20 | 59.50 | 59.15 | 64.88 | 65.25 | 51.30 | 34.40 | 60.20 | 112.51 | 107.50 |
| 21.0 | 60.96 | 60.25 | 59.90 | 65.70 | 66.08 | 51.95 | 34.83 | 60.96 | 113.93 | 108.86 |
| 22.0 | 61.74 | 61.03 | 60.67 | 66.54 | 66.93 | 52.62 | 35.28 | 61.75 | 115.39 | 110.26 |
| 23.0 | 62.55 | 61.82 | 61.45 | 67.41 | 67.80 | 53.30 | 35.75 | 62.55 | 116.89 | 111.69 |
| 24.0 | 63.37 | 62.63 | 62.26 | 68.29 | 68.69 | 54.01 | 36.21 | 63.37 | 118.43 | 113.16 |
| 25.0 | 64.21 | 63.47 | 63.09 | 69.20 | 69.61 | 54.73 | 36.70 | 64.22 | 120.00 | 114.67 |
| 26.0 | 65.08 | 64.32 | 63.95 | 70.14 | 70.55 | 55.47 | 37.20 | 65.09 | 121.62 | 116.22 |
| 27.0 | 65.97 | 65.21 | 64.82 | 71.10 | 71.51 | 56.22 | 37.70 | 65.98 | 123.29 | 117.81 |
| 28.0 | 66.89 | 66.11 | 65.72 | 72.10 | 72.51 | 57.00 | 38.22 | 68.89 | 125.00 | 119.44 |
| 29.0 | 67.83 | 67.04 | 66.65 | 73.10 | 73.53 | 57.80 | 38.77 | 67.84 | 126.77 | 121.13 |
| 30.0 | 68.80 | 68.00 | 67.60 | 74.15 | 74.58 | 58.63 | 39.31 | 68.80 | 128.58 | 122.86 |

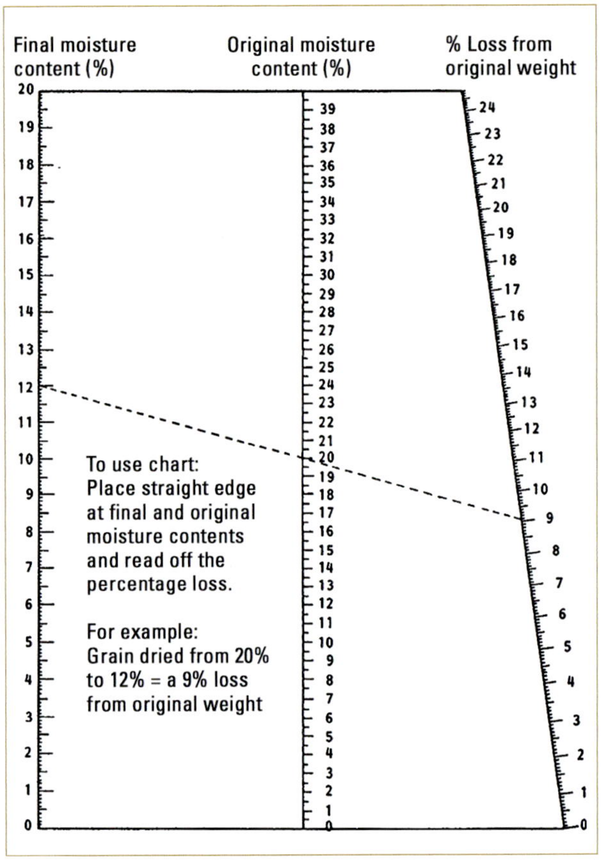


**FIGURE A-5. Estimating weight loss during drying, for grain or seed.**

Chart indicates water loss only. During drying, additional losses (called invisible handling or shrink) are ½ % to 1½ % of original weight. Add additional losses to chart losses. For example, if the chart loss is 9%, then the total loss is about 9½ to 10½ %.

*Source: University of Georgia College of Agriculture Bulletin N.S. 33.*

**TABLE A-15. Approximate water in ear corn based on kernel moisture.**

Values based on No. 2 corn, 15% moisture content, 56 lb/bu for shelled corn and 70 lb/bu for ear corn. Kernel dry matter is 47.3 lb/bu, cob dry matter is 11.4 lb/bu, and total ear corn dry matter is 58.7 lb/bu. Moisture contents are determined on a wet basis.

| Moisture content | | Water in | | |
|---|---|---|---|---|
| Kernel % | Cob % | Kernels lb/bu | Cobs | Ear |
| 10 | 8.6 | 5.3 | 1.1 | 6.3 |
| 12 | 11.7 | 6.5 | 1.5 | 8.0 |
| 14 | 15.5 | 7.7 | 2.1 | 9.8 |
| 15 | 18.6 | 8.7 | 2.6 | 11.3 |
| 16 | 20.0 | 9.0 | 2.9 | 11.9 |
| 18 | 26.2 | 10.4 | 4.1 | 14.4 |
| 20 | 33.8 | 11.8 | 5.8 | 17.7 |
| 22 | 40.0 | 113.4 | 7.6 | 21.0 |
| 24 | 44.3 | 14.9 | 9.1 | 24.0 |
| 26 | 48.0 | 16.6 | 10.5 | 27.2 |
| 28 | 50.6 | 18.4 | 11.7 | 30.1 |
| 30 | 52.8 | 20.3 | 12.8 | 33.0 |
| 35 | 56.4 | 25.5 | 14.7 | 40.2 |

*Source: Principles, Equipment, and Systems for Corn Harvesting, Agricultural. Consulting Associates, Inc., Wooster, Ohio, 1966.*

**EXAMPLE A-4.**

Determine the weight loss as a percentage of original weight when grain is dried from 20% to 12% moisture content.

**SOLUTION A-4.**

Using Figure A-5, weight loss is 9%.

**EXAMPLE A-5.**

Determine the amount of water removed when drying shelled corn from 26% to 15 ½ % moisture content (m.c.).

**SOLUTION A-5.**

From Table A-14, 63.95 lb of 26% m.c. shelled corn is needed to yield a bushel of 15 ½ % m.c. shelled corn (56 lb.). The amount of water removed is:

$$63.95 \text{ lb} - 56.0 \text{ lb} = 7.95 \text{ lb of water}$$

## HARVESTING CAPACITY

To estimate harvesting capacity use Equation A-1:

**EQUATION A-1.**

$$HC = (S \times W \times EFF \times Y) \div 8.25$$

HC = harvesting capacity, bu/hr
S = combine speed, mph
W = combine harvesting width, ft
EFF = harvesting efficiency = 0.65 to 0.85
Y = yield, bu/acre

Harvesting efficiency considers downtime for unloading combines, making adjustments, and turning.

*Use a harvest efficiency of:*

- 0.85 for long fields and no stopping to unload
- 0.75 for long fields and stopping or short fields requiring more turning and no stopping
- 0.65 for short fields and stopping to unload

**EXAMPLE A-6.**

Determine the harvesting capacity for combining corn. The combine travels 4 mph, has a head for four 38-inch rows, and is stopped to unload. Yield is 150 bu/acre.

**SOLUTION A-6.**

1. Assume a harvesting efficiency of 0.75, because harvesting is stopped to unload.

2. Calculate the combine harvesting width.

$$W = 4 \text{ rows} \times 3 \text{ in/row} \div 12 \text{ in/ft} = 12.67 \text{ ft}$$

3. Using Equation A-1, the harvesting capacity is:

$$HC = (4 \text{ mph} \times 12.67 \text{ ft} \times 0.75 \times 150 \text{ bu/A}) \div 8.25 = 691 \text{ bu/hr}$$

## AIR STATIC PRESSURE ▶

## TABLE A-16. Static pressures for airflow through grain.

Derived from ASAE Standard D245.4 1985. For grain only, does not include pressure losses due to duct entrances, transitions or grain-perforated floor interface. Neglect losses for full drying floor with open area greater than 7% and air velocities less than 2,000 fpm in the transition. Use table values directly. For duct or pad type aeration systems and duct velocity less than 2,000 fpm, add ½" of static pressure to the table value. If duct or transition velocity is greater than 2,000 fpm, contact an engineer to determine the static pressure. A multiplier factor is used to convert the pressure drop of clean unpacked grain as given by ASAE Data: ASAE D272 to field conditions.

| Grain type | Grain depth ft | Airflow rate cfm/bu 1/20 | 1/10 | 1/5 | 1/2 | 3/4 | 1 | 1 1/4 | 1 1/2 |
|---|---|---|---|---|---|---|---|---|---|
| | | static pressure, inches of water column | | | | | | | |
| Barley (Multiplier factor = 1.5) | 10 | | | 0.3 | 0.7 | 1.1 | 1.3 | 2.0 | 2.6 |
| | 15 | | 0.3 | 0.6 | 1.6 | 2.7 | 3.8 | 5.0 | 6.4 |
| Oats* | 20 | | 0.5 | 1.1 | 3.0 | 5.0 | 7.3 | 9.8 | 12.4 |
| Sunflower* | 25 | 0.3 | 0.8 | 1.7 | 5.1 | 8.4 | 12.0 | 16.3 | |
| | 30 | 0.6 | 1.2 | 2.5 | 7.6 | 12.7 | | | |
| | 35 | 0.8 | 1.6 | 3.5 | 10.8 | | | | |
| | 40 | 1.0 | 2.2 | 4.7 | 14.6 | | | | |
| | 50 | 1.6 | 3.5 | 7.7 | | | | | |
| | 60 | 2.4 | 5.1 | 11.5 | | | | | |
| | 70 | 3.2 | 7.0 | | | | | | |
| | 80 | 4.3 | 9.5 | | | | | | |
| | 90 | 5.5 | 12.2 | | | | | | |
| | 100 | 6.9 | | | | | | | |
| Shelled corn (Multiplier factor = 1.5) | 10 | | | 0.1 | 0.3 | 0.5 | 0.8 | 1.1 | 1.4 |
| | 15 | | 0.1 | 0.3 | 0.8 | 1.4 | 2.0 | 2.8 | 3.6 |
| | 20 | | 0.2 | 0.5 | 1.6 | 2.7 | 4.1 | 5.6 | 7.4 |
| | 25 | | 0.4 | 0.8 | 2.6 | 4.7 | 7.1 | 9.9 | 12.9 |
| | 30 | | 0.5 | 1.2 | 4.1 | 7.2 | 11.2 | 15.5 | |
| | 35 | 0.4 | 0.8 | 1.7 | 5.9 | 10.6 | | | |
| | 40 | 0.5 | 1.0 | 2.3 | 8.1 | | | | |
| | 50 | 0.7 | 1.6 | 3.9 | 14.1 | | | | |
| | 60 | 1.1 | 2.5 | 5.9 | | | | | |
| | 70 | 1.5 | 3.5 | 8.6 | | | | | |
| | 80 | 2.0 | 4.7 | 11.9 | | | | | |
| | 90 | 2.6 | 6.1 | | | | | | |
| | 100 | 3.3 | 7.8 | | | | | | |
| Soybeans (Multiplier factor = 1.3) | 10 | | | | 0.2 | 0.4 | 0.5 | 0.7 | 0.8 |
| | 15 | | | 0.2 | 0.6 | 0.9 | 1.3 | 1.7 | 2.2 |
| | 20 | | 0.2 | 0.4 | 1.0 | 1.7 | 2.4 | 3.6 | 4.6 |
| | 25 | | 0.3 | 0.6 | 1.7 | 2.8 | 4.2 | 5.7 | 7.5 |
| | 30 | 0.2 | 0.4 | 0.9 | 2.5 | 4.3 | 6.5 | 9.0 | 11.7 |
| | 35 | 0.3 | 0.6 | 1.2 | 3.6 | 6.2 | 9.3 | 13.1 | |
| | 40 | 0.4 | 0.8 | 1.6 | 4.9 | 8.6 | 13.0 | | |
| | 50 | 0.6 | 1.2 | 2.6 | 9.3 | | | | |
| | 60 | 0.8 | 1.7 | 3.9 | 13.7 | | | | |
| | 70 | 1.1 | 2.4 | 5.4 | | | | | |
| | 80 | 1.5 | 3.2 | 7.3 | | | | | |
| | 90 | 1.9 | 4.1 | 9.5 | | | | | |
| | 100 | 2.4 | 5.2 | 12.2 | | | | | |
| Wheat (Multiplier factor = 1.3) | 10 | | | 0.4 | 1.1 | 1.6 | 2.3 | 3.0 | 3.7 |
| | 15 | | 0.5 | 0.9 | 2.5 | 4.0 | 5.6 | 7.2 | 9.2 |
| Grain sorghum* | 20 | | 0.8 | 1.7 | 4.6 | 7.4 | 10.4 | | |
| | 25 | 0.7 | 1.3 | 2.7 | 7.5 | 12.0 | | | |
| | 30 | 1.0 | 1.4 | 3.9 | 11.1 | | | | |
| | 35 | 1.3 | 2.6 | 5.4 | | | | | |
| | 40 | 1.7 | 3.3 | 7.1 | | | | | |
| | 50 | 2.6 | 5.3 | 11.4 | | | | | |
| | 60 | 3.7 | 7.7 | | | | | | |
| | 70 | 5.1 | 10.6 | | | | | | |
| | 80 | 6.7 | 14.1 | | | | | | |
| | 90 | 8.5 | | | | | | | |
| | 100 | 10.7 | | | | | | | |

*Limited research results available for oats, sunflower, and grain sorghum. Values for barley can be used as a good estimate for oats and sunflower. Wheat values can be used for grain sorghum.

# SIZING MOTORS, PULLEYS, AND BELTS

## Motors

Calculate motor horsepower required for conveying equipment by multiplying the hp/10-foot values from the appropriate tables in this handbook (Tables 3-3, 3-7, 3-8) by the conveyor length/10-foot value. Next choose the motor horsepower from Table A-17. See Example 3-2.

Electric motors mounted with the shaft in positions other than horizontal must be designed for the bearings to support the motor under these conditions. Operate ½ horsepower and larger electric motors on 240-volt service (nominal) and control them with a magnetic motor-starting switch. Attach the equipment grounding wire to the electric motor frame for shock protection. Make sure the conductor can carry at least 125% of the motor nameplate current. Make sure overcurrent protection is properly sized for the motor and circuit. If an on-off switch is used for small motors, install time delay fusetrons or fusestats sized to 115% of the nameplate operating current.

**TABLE A-17. Motor selection for continuous by operating conveyor.**

| Calculated conveyor hp | Electric motor hp | Gasoline engine hp |
|---|---|---|
| Up to 0.27 | 1/4 | 1/2 |
| 0.28 to 0.35 | 1/3 | 2/3 |
| 0.36 to 0.55 | 1/2 | 1 |
| 0.56 to 0.81 | 3/4 | 1 ½ |
| 0.82 to 1.10 | 1 | 2 |
| 1.11 to 1.60 | 1 ½ | 3 |
| 1.61 to 2.10 | 2 | 4 |
| 2.11 to 3.20 | 3 | 5 |
| 3.21 to 5.25 | 5 | 8 |
| 5.26 to 7.90 | 7 ½ | 12 |
| 7.91 to 10.50 | 10 | 15 |

## Pulleys

Most electric drive motors used to power conveyers run at 1,725 rpm. Gas engines run at relatively higher rpm. Usually, equipment needs to run more slowly than the motor. Use different size pulleys on motors and equipment to adjust speed. The motor pulley is usually smaller than the equipment pulley. For electric motors ½ horsepower or larger, use at least a 3-inch outside diameter motor pulley.

To calculate equipment pulley size, choose a pulley size for the motor first. Use Equation A-2 to calculate equipment pulley size. Pulley size calculations are based on pitch diameter, not outside diameter, Figure A-6.

**EQUATION A-2.**

$$EP = (MRPM \times MP) \div ERPM$$

EP = equipment pulley pitch diameter, in.
MRPM = motor rpm, revolutions per minute, usually 1.725 rpm but check the motor nameplate.
MP = motor pulley pitch diameter, in.
ERPM = desired equipment rpm

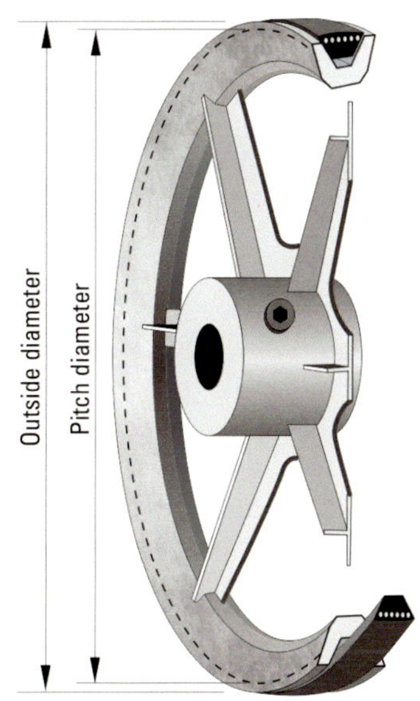


Pitch diameter for A-belt = outside diameter - 0.22"
Pitch diameter for B-belt = outside diameter - 0.38"
Pitch diameter for A-belt in a B-belt pulley = outside diameter - 0.78"

**FIGURE A-6. Outside and pitch diameter of a pulley.**

## Belts

V-belts are classified according to cross-sectional size, shown in Figure A-7. Belt type and the number of belts required for a piece of equipment depends on the horsepower being transmitted, Table A-18. If belts are too small or not enough belts are used, they will slip and wear out quickly. Select pulleys with the same width as the belt or belts, illustrated in Figure A-7. This is particularly important if equipment is to operate at the calculated speed.

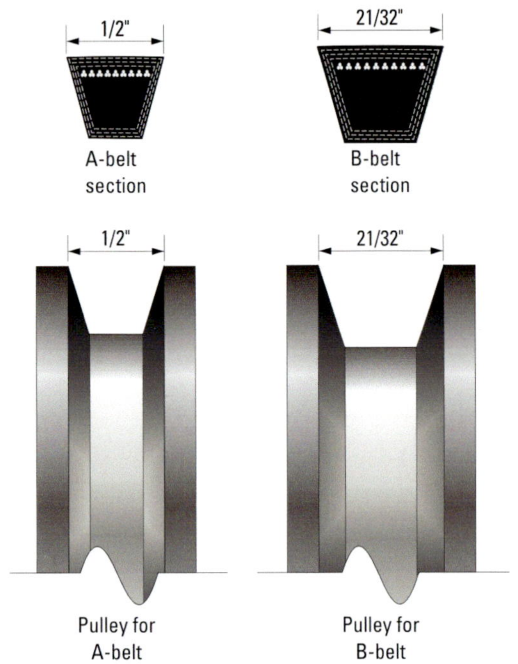


**FIGURE A-7. Typical V-belts and pulleys for grain equipment.**

### EXAMPLE A-7.

A 6-inch auger has a 5 hp motor with a 4-inch pulley that operates at 1,725 rpm. Determine the belts and auger pulley size required to operate at 600 rpm.

### SOLUTION A-7.

1. From Table A-18, three A-belts are required to transmit 5 hp with a 4-inch motor pulley.

2. Calculate the motor pulley pitch diameter. From Figure A-6, subtract 0.22 from the outside diameter.

$$\text{Pitch diameter} = 4 \text{ in} - 0.22 \text{ in} = 3.78 \text{ in}$$

3. Calculate the auger pulley pitch diameter using Equation A-2.

$$EP = (1{,}725 \text{ rpm} \times 3.78 \text{ in}) \div 600 \text{ rpm} = 10.87 \text{ in}$$

4. The outside diameter of the auger pulley is:

$$10.87 \text{ in} + 0.22 \text{ in} = 11.09 \text{ in}$$

Select an 11-inch auger pulley.

**TABLE A-18. Required number and type of V-belts.**

When more than 3 belts are required, consider a chain drive.

| Motor pulley outside dia., in | Motor hp | | | | | | | | |
|---|---|---|---|---|---|---|---|---|---|
| | ½ or under | 3/4 | 1 | 1 ½ | 2 | 3 | 5 | 7 ½ | 10 |
| | Number and type of belt(s) required | | | | | | | | |
| 3 | 1-A | 1-A | 1-A | 2-A | 2-A | 3-A | 5-A | | |
| 3 ½ | 1-A | 1-A | 1-A | 2-A | 2-A | 3-A | 4-A | | |
| 4 | 1-A | 1-A | 1-A | 1-A | 2-A | 2-A | 3-A | 5-A | |
| 4 ½ | 1-A | 1-A | 1-A | 1-A | 1-A | 2-A | 3-A | 5-A | 5-A |
| 5 | 1-A | 1-A | 1-A | 1-A | 1-A | 2-A | 3-A | 4-A | 5-A |
| 5 ½ | 1-A | 1-A | 1-A | 1-A | 1-A | 1-B | 2-B | 3-B | 4-B |
| 6 | 1-A | 1-A | 1-A | 1-A | 1-A | 1-B | 2-B | 2-B | 3-B |
| 7 | 1-A | 1-A | 1-A | 1-A | 1-A | 1-B | 2-B | 2-B | 3-B |
| 8 | 1-A | 1-A | 1-A | 1-A | 1-A | 1-B | 1-B | 2-B | 2-B |

# CONVERSION FACTORS FOR GRAIN SYSTEMS

**TABLE A-19. Bulk grain handling units.**

| | English | | Metric | |
|---|---|---|---|---|
| Physical quantity | Unit | Symbol | Unit | Symbol |
| Length | foot | ft or ' | meter | m |
| Length | inch | in or " | centimeter or millimeter | cm or mm |
| Weight | pound | lb | kilogram | kg |
| Land area | acre | acre | hectare | ha |
| Volume | cubic foot | $ft^3$ | cubic meter | $m^3$ |
| Flowrate | cubic feet per minute | $ft^3/min$ | cubic meter per second | $m^3/s$ |
| Static pressure | inches of water column | in | kilopascal | kPa |
| Crop yield | bushel* per acre | bu/acre | kilograms per hectare | kg/ha |
| Production | bushel | bu | metric ton | t or Mg |
| Unit of trade | bushel | bu | metric ton | t or Mg |

*dry matter bushel

**TABLE A-20. Conversions.**

Multiply to the right: acres × 43.560 = $ft^2$.
Divide to the left: $ft^2$ ÷ 43,560 = acres.

| Unit | Times | Equals |
|---|---|---|
| **Acre** | 43,560 | $ft^2$ |
| | 4,840 | $yd^2$ |
| | 160 | square rods |
| | 1/640 | square mile |
| **Bushel**** | 1.245 | $ft^3$ |
| | 2.5 | $ft^3$ ear corn |
| **Cubic feet, $ft^3$** | 1728 | $in^3$ |
| | 0.4 | bu ear corn |
| | 0.8 | bu grain (volume) |
| **Cubic yard** | 27 | $ft^3$ |
| concrete | 81 | $ft^2$ of 4" floor |
| concrete | 54 | $ft^2$ of 6" floor |
| **Mile** | 5,280 | ft |
| | 1,760 | yd |
| | 320 | rods |
| **Pressure, psi** | 2.31 | ft of water column |
| | 27.69 | in. of water column |
| **Rod** | 16.5 | ft |
| | 5.5 | yd |
| **Horsepower** | 0.745 | kW |

** volume bushel

**TABLE A-21. Conversions.**

Multiply to the right: inches × 2.54 = cm.
Divide to the left: cm ÷ 2.54 = inches.

| Unit | Times | Equals |
|---|---|---|
| **Length** | | |
| inch | 25.40 | mm |
| foot | 0.3048 | m |
| yard | 0.9144 | m |
| mile | 1.609 | km |
| **Area** | | |
| $in^2$ | 6.451 | $cm^2$ |
| $ft^2$ | 0.09290 | $m^2$ |
| $yd^2$ | 0.8361 | $m^2$ |
| acres | 0.4047 | ha |
| acres | 4047.0 | $m^2$ |
| $mile^2$ | 2.590 | $km^2$ |
| **Volume** | | |
| $in^3$ | 16.39 | $cm^3$ |
| $in^3$ | 0.01639 | liters (L) |
| $ft^3$ | 0.02832 | $m^3$ |
| cubic yards | 0.7676 | $m^3$ |
| bushel | 0.03524 | $m^3$ |
| **Mass or weight** | | |
| pounds | 0.4536 | kg |
| ton (2,000 lb) | 907.2 | kg |
| ton (2,000 lb) | 0.9072 | tonne (t) |
| **Velocity or speed** | | |
| ft/sec | 0.3048 | m/s |
| miles/hour (mph) | 1.609 | km/h |
| **Flowrate** | | |
| cfm | 0.0004719 | $m^3/s$ |
| **Pressure** | | |
| in of water column | 0.2488 | kPa |
| **Temperature** | | |

Fahrenheit (F) = (C × 1.8) + 32
Celsius (C) = (F – 32) ÷ 1.8

| TABLE A-22. Grain production conversions. | | |
|---|---|---|
| A volume bushel is 1.245 ft$^3$. Because different grains have different weight bushel definitions, a special conversion factor is used for each crop. Values based on lb/bu from Table A-4 and 2,200 lb/metric ton. | | |
| | **Weight bushels to metric tons** | **Metric tons to weight bushels** |
| **Grain** | **Multiply by** | |
| Barley | 0.0218 | 45.930 |
| Corn | 0.0254 | 39.368 |
| Flaxseed | 0.0255 | 39.286 |
| Oats | 0.0145 | 68.750 |
| Rapeseed | 0.0227 | 44.000 |
| Rye | 0.0255 | 39.286 |
| Sorghum | 0.0254 | 39.368 |
| Soybeans | 0.0272 | 36.744 |
| Wheat | 0.0272 | 36.744 |

| TABLE A-23. Yield conversions. | | |
|---|---|---|
| | **Dry matter bu/ac to kg/ha** | **kg/ha to bu/ac dry matter** |
| **Grain** | **Multiply by** | |
| Barley | 53.81 | 0.0186 |
| Corn | 62.78 | 0.0159 |
| Flaxseed | 62.78 | 0.0159 |
| Oats | 35.87 | 0.0279 |
| Rapeseed | 56.05 | 0.0178 |
| Rye | 62.78 | 0.0159 |
| Sorghum | 67.26 | 0.0159 |
| Soybeans | 67.26 | 0.0149 |
| Wheat | 67.26 | 0.0149 |

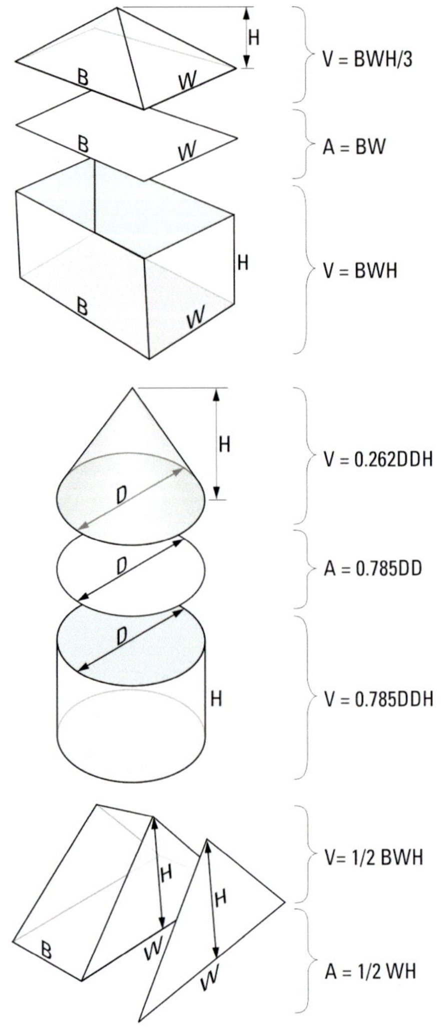


**FIGURE A-8. Areas and volumes.**

# GLOSSARY

A

**Aeration:** Process of moving air through stored dry grain to control grain temperature.

**Airflow:** The volume of air moved per unit time L/s, cfm.

**Airflow rate:** The volume of air moved per unit of time and crop, cfm/bu

**Air lock:** A device that meters grain into the airstream of a pneumatic conveyor while forming an air seal to prevent airflow through the air lock.

**Air velocity, apparent (feet per minute):** The rate of airflow determined by dividing the volume of airflow by the cross-sectional area.

**Allowable storage time:** The amount of time a product can be held before respiration and mold growth cause enough deterioration to result in a loss in commercial grade and value. For shelled corn, it is defined as the amount of time before there is a 0.5% loss of the initial amount of dry matter.

**Ambient temperature:** The dry bulb temperature of outdoor air.

**AMCA:** Air Movement and Control Association; an independent organization responsible for fan testing, specifications, and procedures.

**Auger or screw conveyor:** A device that moves granular products by forcing them along a rotating screw mounted inside a fixed tube or trough.

**Auger, sweep:** An auger without a tube mounted horizontally above the floor of a bin to move stored product to the center sump during unloading.

**Auger, underfloor (unloading):** An under-floor screw conveyor used to move grain from center sump to outside of the bin.

**Automatic/Automation:** A process (i.e.; drying or conveying) controlled without human action.

B

**Batch:** A quantity of harvested crop put into a bin or container on a repetitive basis for treatment, such as drying.

**BCFM:** Broken corn and foreign material for shelled corn. Generally referred to as fines in this book.

**Bin:** A structure for storing grain or other bulk material. Also called a silo.

**Bin, holding:** A bin (usually with a hopper bottom) used to hold grain before it is dried or loaded out.

**Bin, hopper-bottom:** A self-cleaning bin that has a bottom with sloping sides so that product flows by gravity to the unloading port.

**Bin, surge:** A bin to accumulate and temporarily hold grain before being conveyed.

**Biomass:** Plant or organic material used as a fuel or other bioproduct.

**Bollard:** A short vertical post designed to protect an area from vehicular traffic.

**Bucket elevator (leg):** A vertical conveyor with a large number of small open-topped containers (buckets) attached to a continuous belt, used for raising grain from one elevation to another.

**Bushel, dry matter:** One bushel = a standard weight at a standard moisture content (dry matter bushel). This definition corrects for moisture content by adjusting weights to an equivalent weight of dry matter. Dry matter bushels are sometimes referred to as "equivalent bushel" and are used in reporting yield and in research, but not for trade.

**Bushel, volume:** One bushel = 1.245 $ft^3$ (volume, or Winchester, bushel). This is the bushel used for the quality characteristic test weight. Often 1.245 $ft^3$ is rounded to 1.25 $ft^3$. Volume bushels are used to determine the storage capacity of bins, trucks, etc.

**Bushel, weight:** One bushel = a standard weight at any moisture content (weight bushel). This is the bushel definition used by the grain trade. Every grain has its own standard weight.

**bu/hr or bu/in:** bushels per hour

C

**Cash grain:** Grain that is grown and processed for sale rather than stored and fed to livestock or poultry.

**cfm:** Cubic feet per minute ($ft^3$/min); unit of airflow.

**cfm per bu:** Cubic feet per minute per volume bushel; unit of airflow per unit of volume.

**Cleaner/screener:** A device used to remove undesirable materials or particle sizes from a product.

**Cleaner, air-screen:** A device that separates seeds on the basis of size, shape, and density using screens and airflow.

**Cleaner, rotary:** A device that separates using screens rotating about a central axis.

**Conveyor, bulk or mass flow:** An enclosed conveyor with shallow paddles or flights attached to a chain or cable. The shallow flight scraping along the housing bottom carries grain 3 to 5 times as deep as the flight.

**Conveyor, drag:** An enclosed conveyor that has a number of paddles or flights attached to a chain and drags product. Also referred to as a mass flow conveyor.

**Conveyor, flight:** An enclosed or open conveyor with paddles or flights attached to one or two chains.

**Cooling bin:** Separate bin used to cool grain dried with a high temperature dryer.

**Cushion box:** A device installed in a downspout that fills with grain to reduce impact damage to grain.

D

**Distributor:** A device that rotates to divert product from a single central conveyor, generally a bucket elevator, to one of several different down spouts or discharges.

**Downspout:** An enclosed tube that directs product from a bucket elevator or overhead conveyor to the desired location.

**D-ring (for lifelines):** A D-shaped metal loop used to connect a lifeline to a body harness or anchor point.

**Dryeration:** The process of cooling grain from a high temperature dryer after the grain has steeped for at least 4 hours which results in moisture removal due to evaporative cooling.

**Duct:** Circular or rectangular tube that air flows or moves through in a drying or aeration system.

E

**Effective elevating height:** Vertical distance from the base of a bucket elevator boot or transport auger to the discharge height, which determines power requirements for the conveyor motor.

F

**F:** Fahrenheit; °F (degree F) is a unit of temperature

**Fan, axial-flow:** Fan in which air flows parallel to the impeller's axis of rotation; common types are tube-axial and vane-axial.

**Fan, centrifugal:** A fan in which air enters parallel to the impeller's axis of rotation, moves radially through the blades and is discharged tangentially from a scroll housing surrounding the impeller. Common types used are low-speed, high-speed, and in-line centrifugal fans.

**Fan, centrifugal–high speed:** A centrifugal fan that operates at about 3,500 rpm.

**Fan, centrifugal–in-line:** A centrifugal fan mounted in a tube shaped housing with a bell shaped inlet directing the air into the center of the impellor. The fan typically operates at about 3,500 rpm. Also referred to as a tubular centrifugal fan.

**Fan, centrifugal–low speed:** A centrifugal fan that operates at about 1,750 rpm.

**Fan, vane-axial:** A tube-axial fan that uses guide vanes inside the housing to help straighten air to reduce turbulence.

**Fan efficiency:** The volumetric airflow per unit of input power, cfm/W.

**Fines/Dockage:** Broken kernels, foreign material, cobs, weed seeds, etc. intermixed in the grain. Fines refers to corn; dockage to small grains.

**FM:** Foreign material

**fpm:** feet per minute (ft/min); unit of velocity or speed.

**ft:** feet, unit of length.

**$ft^2$:** square feet; unit of area.

**$ft^3$:** cubic feet; unit of volume.

G

**gal:** gallon

**General Duty Clause (OSHA):** Section 5(a)(1) of the Occupational Safety and Health Act of 1970 that reads, "Each employer shall furnish to each of his employees employment and a place of employment that are free from recognized hazards that are causing or are likely to cause death or serious physical harm to his employees."

**Grain center:** The complete grain drying, handling, and storage system. Includes equipment or structures for grain receiving, conveying, drying, cooling, storage, and loadout.

**Grain conditioning:** Changing the temperature and/or moisture of grains to achieve desired values.

**Gravity flow:** Non-mechanical flow; results from a difference in elevation between the discharge of a holding bin and desired location receiving the grain or the inlet and outlet of a bucket elevator transfer tube.

**Gravity spouting discharge height:** Vertical distance from the bucket elevator distributor discharge, valves, or grain cleaner to the point of delivery.

H

**High temperature (high speed) dryer:** A dryer that dries grain relatively fast with air heated to at least 120°F.

**Hopper**: The sloped portion of a bin that is used to aid gravity discharge through an orifice.

**hp:** horsepower; unit of power.

**Hub:** The grain center receiving and/or loadout area.

**Humidity, relative**: A measure of moisture in air expressed as a percentage. It is the ratio of the weight of water vapor in a given volume of dry air at a given temperature to the maximum quantity of water vapor that the same volume of air could hold at that temperature.

I

**Impeller (blades)**: The rotating part of a fan that causes air to move.

**in:** inches; unit of length

K

**Knot-passing pulley:** A suspended pulley block with side plates that allow knots to pass between the plates and over the pulley; often used with lifelines.

L

**Leg:** Common name for a bucket elevator.

**Low temperature dryer:** A dryer that slowly dries grain with unheated air or air heated up to 10°F.

**Lifeline (grain bin entry):** A properly anchored and controlled line connected to a worker wearing a full-body harness inside a bin, to prevent the worker from becoming entrapped in grain.

**Lockout/tagout:** The procedure of shutting off a machine or other hazard prior to working on or near it, and using a lock to prevent it from being restarted or automatically starting. The tag identifies the worker who applied the lock; only that worker should be able to remove it.

**LP gas:** Liquefied petroleum gas.

M

**Management center building:** A building that houses controls, testing equipment, and records for grain drying, handling, and storage.

**Manometer:** Instrument for measuring static pressure (resistance to airflow).

**Metric tonne:** Metric unit of weight; 1 metric tonne = 2,200 lb.

**Moisture content:** (w.b.–wet basis): Weight of water in the product divided by the weight of wet material (total weight), expressed as percentage. This value is normally used in agricultural marketing.

**Negative pressure system:** Refers to an air distribution system in which the pressures in the air distribution ducts and plenums are below atmospheric pressure; air is typically sucked through the product toward the fan.

P

**Pack factor:** A factor used to take into account the increase in static pressure or grain resistance to air flow commonly found in bins. This is not to be confused with the packing factor used to determine storage capacity of bins and other storage structures.

**Pit:** An open-topped underground container that is covered with a grate and has sides sloping to a conveyor at its bottom. Pits serve as surge containers that allow transport vehicles to unload product faster than conveyors can move it to drying or storage facilities.

**Positive pressure system:** Refers to an air distribution system in which the pressures in the air distribution ducts and plenums are above atmospheric pressure; air is typically pushed through the product away from the fan.

**PTO:** Power Take-Off. System to obtain power from a tractor through a rotating shaft.

R

**rpm:** Revolutions per minute—usually refers to the rotating speed of shafts, pulleys, and augers.

S

**Screenings:** Fines and other material separated from grain in the cleaning process.

**Self-contained dryer:** A unit that dries and cools grain or transfers hot grain. Has little or no wet holding capacity. Most often the same as a high temperature (high speed) dryer.

**Spoutline:** The vertical column directly under the grain delivery point (spout) used to fill a storage. There is normally a higher concentration of fines in the spoutline.

**Silo:** A structure for storing grain or other bulk material. Also called a bin.

**Static efficiency:** Fan efficiency calculated by dividing static air power (based on static pressure) by power output of the fan motor.

**Static pressure:** Resistance that needs to be overcome to move air through a grain system. Usually measured in inches of water column.

**Steep:** To hold hot grain taken directly from a high temperature dryer for 4 to 12 hours to permit the moisture and temperature to equalize within the grain. Used with in-storage cooling and dryeration to increase grain quality, increase drying capacity, and lower drying costs.

T

**Test weight:** Test weight is a measure of grain bulk density in units of pounds per bushel, where a bushel is 1.245 $ft^3$. Test weight is a general indicator of grain quality and higher test weight normally means higher quality grain. It is not used as the basis for payment of grain.

**Transition:** Usually a tapered duct between a fan and distribution duct or the perforated floor drying/ storage bin

# REFERENCES


**ANSI.** 2007. ANSI Z359.1-2007. *Safety Requirements for personal fall arrest systems, subsystems, and components.* American National Standards Institute, New York, NY.

**ASABE.** 2014. ASAE EP344.4 JAN2014. *Lighting Systems for Agricultural Facilities.* American Society of Agricultural and Biological Engineers, St. Joseph, MI

**ASABE.** 2014. ASAE S412.1 MAR1990 (R2014). *Ladders, Cages, Walkways and Stairs.* American Society of Agricultural and Biological Engineers, St. Joseph, MI.

**ASSE.** 2007. Special issue on the new ANSI/ASSE Z359 fall protection code. *By Design, a technical publication of ASSE's Engineering Practice Specialty.* American Society of Safety Engineers, Des Plaines, IL. www.asse.org/assets/1/7/ByDesign_Z359Special_Fall2007.pdf Downloaded 5-23-2016.

**Bauer, W. and B. Harp.** 2014. Grain Entrapment Prevention Systems. Presentation at GEAPS Exchange, Omaha, NE, 2-24-2014. GEAPS (The Grain Elevator and Processing Society), Golden Valley, MN. www.gfai.org/docs/Tues_Grain_Bin_Redesign.pdf Downloaded 5-23-16.

**Bauer, W.** 2013. *What is a secured lifeline?* Grain Journal 41(6): 56-59.

**Bauer, W.** 2014. *Grain bin anchors.* Grain Journal 42(1): 58-64.

**Bauer, W.** 2011. *How to properly address engulfment hazards.* Presentation at the Grain Entrapment Prevention Symposium, March 23-24, 2011, Assumption, IL http://grainnet.com/pdf/C_GEAP_3_23_2011.pdf Downloaded 5-23-16

**Carpenter, R.** 2012. *Tunnels, Reclaim Systems & Soil Borings.* Presentation at the 2012 Grain Entrapment Prevention Symposium, March 21-13, 2012. St. Louis, MO. Clear Creek & Associates, Goshen, IN. http://www.grainnet.com/pdf/Rodney_Carpenter.pdf Downloaded 5-23-16

**Grain Bin Entrapment Initiative.** Seven Grain Entrapment Prevention Principles. http://www.grainnet.com/articles/seven_grain_entrapment_prevention_principles-135048.html Accessed 5-23-16

**Grain Handling Safety Coalition.** 2014. *Learn and Live: Grain Bin Entry module.* Produced under grant number SH-22288SH1 from the Occupational Safety and Health Administration, U. S. Department of Labor. Grain Handling Safety Coalition, Springfield, IL. http://grainsafety.org/wp-content/uploads/2014/08/GrainBinEntryFinalPDFJan2014.pdf Downloaded 5-23-16.

**Grain Handling Safety Coalition.** 2013. *Learn and Live: Lifeline System for Bin Entry mini-module.* Produced under grant number SH-22288SH1 from the Occupational Safety and Health Administration, U. S. Department of Labor. Grain Handling Safety Coalition, Springfield, IL. http://grainsafety.org/training-2/life-line-protection-system/ Downloaded 5-23-16

**Hellevang, K.** 1993 (revised 2005), EB-35 Natural Air/Low Temperature Crop Drying https://www.ag.ndsu.edu/graindrying/documents/eb35.pdf

**Lovell, G.** 2013. *Why do we need anchor points in bins?* Presentation at the 2013 Grain Entrapment Prevention Symposium, March 12-13, Lincoln, NE. Safety and Technical Rescue Association (SATRA) Powerpoint presentation titled Grain Bin Safety Awareness http://grainnet.com/pdf/SATRA_Grain_Bin_Safety_Awareness_GeorgeLovell.pdf Downloaded 5-23-16

**Miller and Honeywell.** 2012. *Understanding the ANSI Z359 Fall Protection Code.* Miller Fall Protection, Franklin, PA. https://www.millerfallprotection.com/pdfs/ANSI-Z359Standard/UnderstdgANSI-8-2012.pdf Downloaded 5-23-2016.

**MWPS.** 2004. *Managing Dry Grain in Storage*, AED-20. Prepared by Bruce A. McKenzie extension agricultural engineer, Purdue University, and Larry Van Fossen, extension agricultural engineer, Iowa State University. The sections relating to insect control were prepared by David E. Foster and Harold J. Stockdale, Iowa State Unviersity extension entomologits. www-mwps.sws.iastate.edu/catalog/grain-handling-storage/managing-dry-grain-storage Accessed 5-23-16

**MWPS.** 1999. *Dry Grain Aeration SystemsDesign Handbook*, MWPS-29. Midwest Plan Service, Ames IA 50011. www-mwps.sws.

iastate.edu/catalog/grain-handling-storage/dry-grain-aeration-systems-design-handbook

**MWPS.** 1994. *Farm and Home Concrete Handbook,* MWPS-35. Midwest Plan Service, Ames IA 50011. www-mwps.sws.iastate.edu/catalog/construction-farm/farm-and-home-contrete-handbook

**MWPS.** 2013. *Wiring Handbook for Rural Facilities.* MWPS-28. Midwest Plan Service, Ames IA 50011. www-mwps.sws.iastate.edu/catalog/construction-farm/wiring-handbook-rural-facilities

**Night Work Safety.** Agsafe, Modesto, CA. Agsafe.org http://www.agsafe.org/index.php/en/resourcecenter/safetyandhrresources.html Downloaded 5-24-16

**Oklahoma State University.** Stored Products Research and Education Center—numerous resources: http://sprec.okstate.edu/safety Accessed 5-23-16

**OSHA 1926 Subpart D, 1926.56.** *Illumination.* (Minimum illumination intensities for construction and related areas)

**OSHA 1910.272**—*Grain Handling Facilities*. Within OSHA Subpart R, Special Industries;

**OSHA 1910.23**—*Guarding floor and wall openings and holes* (mentions railings for platforms, floors, and stairs.) Within OSHA Subpart D – Walking-Working Surfaces.

**OSHA 1926 Subpart M**—*Fall Protection*. Wide variety of guidance and requirements.

**OSHA 1910.38**—*Emergency Action Plans*. Within Subpart E, Means of Egress.

**OSHA 1904**—*Recording and Reporting Occupational Injuries and Illnesses*.

**Pneumatic Systems.** Product literature on BinWhip and BinDrill remote bin cleanup and bridging solutions. Pneumat Systems, Mankato, IL http://pneumat.com/resources/literature/ Accessed 5-26-16. (No endorsement of this company or its products is made or implied.)

**Safe Lighting Practices in the Shipyard Industry**. OSHA Fact Sheet. http://www.osha.gov/Publications/osha3677.pdf Downloaded 5-24-16.

**Schwab, C.V.** 1982. *Inflow velocity and forces acting on a person trapped in enveloping flow of granular material.* Unpublished Master's Thesis. Department of Agricultural Engineering, University of Kentucky, Lexington, KY.

**Schwab, C.V., I.J Ross, L.R. Piercy, and B.A. McKenzie.** 1985. *Vertical pull and immersion velocity of mannequins trapped in enveloping grain flow.* Transactions of the ASAE 28(6): 1997-2002. American Society of Agricultural and Biological Engineers, St. Joseph, MI.

**Sukup Manufacturing.** 2016. **Bin Operation Manual—**Commercial, Stiffened, Farm & Hopper. Page 15, Knot-passing pulley and restraint anchor. Manual L13920 © Sukup Manufacturing Co., Sheffield, IA. https://www.sukup.com/content/safety/operation/L13920_BinOperation.pdf Downloaded 5-16-16. (No endorsement of this company or its products is made or implied.)

# RESOURCES

**Iowa State University Extension and Outreach, Iowa Grain Quality Initiative**—http://www.iowagrain.org

**North Dakota State University, Agriculture and University Extension, Grain Drying and Storage**—https://www.ag.ndsu.edu/graindrying

**United States Department of Labor, Safety and Health Topics for Grain Handling**—https://www.osha.gov/SLTC/grainhandling/

**University of Kentucky spreadsheets for calculating grain capacities in various types of storage**—www-mwps.sws.iastate.edu/spreadsheets

**University of Minnesota Extension, Safe Storage and Handling of Grain**—http://extension.missouri.edu/p/G1969

**University of Minnesota Fan Selection for Grain Bins (FANS online program)**—http://webapps.bbe.umn.edu/fans/

# INDEX